Springer Theses

Recognizing Outstanding Ph.D. Research

For further volumes:
http://www.springer.com/series/8790

Aims and Scope

The series "Springer Theses" brings together a selection of the very best Ph.D. theses from around the world and across the physical sciences. Nominated and endorsed by two recognized specialists, each published volume has been selected for its scientific excellence and the high impact of its contents for the pertinent field of research. For greater accessibility to non-specialists, the published versions include an extended introduction, as well as a foreword by the student's supervisor explaining the special relevance of the work for the field. As a whole, the series will provide a valuable resource both for newcomers to the research fields described, and for other scientists seeking detailed background information on special questions. Finally, it provides an accredited documentation of the valuable contributions made by today's younger generation of scientists.

Theses are accepted into the series by invited nomination only and must fulfill all of the following criteria

- They must be written in good English.
- The topic should fall within the confines of Chemistry, Physics, Earth Sciences, Engineering, and related interdisciplinary fields, such as Materials, Nanoscience, Chemical Engineering, Complex Systems, and Biophysics.
- The work reported in the thesis must represent a significant scientific advance.
- If the thesis includes previously published material, permission to reproduce this must be gained from the respective copyright holder.
- They must have been examined and passed during the 12 months prior to nomination.
- Each thesis should include a foreword by the supervisor outlining the significance of its content.
- The theses should have a clearly defined structure including an introduction accessible to scientists not expert in that particular field.

Azadeh Ramesh

Response of Flood Events to Land Use and Climate Change

Analyzed by Hydrological and Statistical Modeling in Barcelonnette, France

Doctoral Thesis accepted by
the University of Vienna, Austria

Springer

Author
Dr. Azadeh Ramesh
Institute of Geomorphological Systems and Risk Research
Department of Geography and Regional Research
University of Vienna
Vienna
Austria

Supervisor
Prof. Dr. Thomas Glade
Institute of Geomorphological Systems and Risk Research
Department of Geography and Regional Research
University of Vienna
Vienna
Austria

ISSN 2190-5053 ISSN 2190-5061 (electronic)
ISBN 978-94-007-5526-0 ISBN 978-94-007-5527-7 (eBook)
DOI 10.1007/978-94-007-5527-7
Springer Dordrecht Heidelberg New York London

Library of Congress Control Number: 2012952039

Printed on acid-free paper

Springer is part of Springer Science+Business Media (www.springer.com)

I dedicate this work to "Alireza"

Parts of this research have been published or prepared for publication at the following conferences and in the following journals:

- Ramesh, A., Glade, T., Malet, J.-P., Trend analysis in hydro-climate parameters of a mountainous catchment in south east of France using Mann-Kendal test, International Journal of Climatology (in prep.).

- Ramesh, A., Alkema, D., Glade, T., Flood hazard assessment using SOBEK (1D-2D) model, Case study: Reconstruction of 1957 and 2008 floods in Barcelonnette, France, International Journal of Hydrology and Earth System Sciences (in prep.).

- Ramesh, A., Glade, T., The response of flood event to land use and climate change in Barcelonnette, France, International Journal of Natural Hazards (in prep.).

- Ramesh, A., Glade, T., Alkema, D., Krol, B., Malet, J.-P., Model Performance Analysis for Flood Hazard Assessment in Ubaye River, Barcelonnette, France, International Conference; Mountain Risk—Bringing Science to Society, CERG Publication, Strasbourg, France, November 2010.

- Ramesh, A., Glade, T., Response of flood event to land use and climate change, International Summer School on Water Resources and the Water Cycle in a Changing World, Oxford, UK, 04–08 July 2011.

- Ramesh, A., Glade, T., Trend analysis in climatic parameters to investigate flow regime of Ubaye River in Barcelonnette, European Conference on Applied Climatology, EMS Annual Meeting, Berlin, Germany, 12–16 September 2011.

- Ramesh, A., Glade, T., Parodi, G., Alkema, D., Dyke failure scenarios in Ubaye River, France using hydraulic modeling, European Geosciences Union, Vienna, Austria, 03–08 April 2011.

- Ramesh, A., Glade, T., Alkema, D., Krol, B., Malet, J.-P., Model performance analysis for flood hazard assessment in Ubaye River, Barcelonnette, France, International Conference; Mountain Risks: Bringing Science to Society, Firenze, Italy, 24–26 November 2010.

- Ramesh, A., Glade, T., Malet, J.-P., Evaluation of climate change on flood event by using parametric T-test and non-parametric Mann-Kendall test in Barcelonnette basin, France, European Conference on Applied Climatology, EMS Annual Meeting, Zurich, Switzerland, 13–17 September 2010.

- Ramesh, A., Glade, T., Kappes, M., Malet, J.-P., Delineation of flood inundated areas using aerial photo interpretation and GIS-based hydrological modeling—an application in Barcelonnette, France, European Geosciences Union, Vienna, Austria, 02–07 May 2010.

Supervisor's Foreword

Floods are one of the most important hydrogeomorphic processes operating on the earth's physical landscape that greatly influences society in manifold ways. These events have been a major concern for people living near rivers since pre-historic times. Floods cause not only one of the most widespread natural disasters often associated with large numbers of casualties, rising economic losses, extensive homelessness, and disaster induced diseases, they are also one of the most frequent disaster type. The Commission to the Council (the European Parliament) reported in 2004, that besides natural causes, river floods may also occur whenever the capacity of man-made drainage system (such as dykes) is unable to cope with the volume of water generated by rainfall, or when flood defences fail. Such failures are one of the most common causes for river flooding in Europe. Apart from non-climatic factors that affect flooding, there are adverse flood hazard changes due to climate change. Many investigators believe that climate change is expected to have substantial impacts on hydrology on a global, regional, and local scale. Unfortunately, a flood is in many locations not a purely natural phenomenon anymore. Some human activities, such as increasing human settlements in flood-plains, increasing water storage through dams, or reducing natural water retention by land use and climate change, contribute to the increased likelihood of flood events. Therefore, studies on floods are in the research focus for many decades. Numerous disciplines are involved and respective topics include past information on floods using geoarchives or historical sources, field mapping and monitoring, 1D, 2D, and 3D modeling approaches as well as any socioeconomic consequences and disaster management in their broadest sense. In recent years, the effects of global change including, for example, land use and climate change receive increased attention. As described in the previous paragraph, it is obvious, that the fluvial system is not only driven by natural forces, it is also hugely determined by societal interactions. Numerous scientific questions associated to this topics are still unsolved, leaving space for further research. Consequently, Azadeh Ramesh dedicates her research to this increasingly important field of research. The given different data sets are analysed with respect to climate change focusing on both temperature and precipitation variations. The land use change analysis is based on aerial photograph imagery to determine spatiotemporal land use patterns. The book's main focus is on flood hazard assessment which is based on hydrological, spatial, and statistical modeling. The study intends to develop an advanced understanding of how climate change and land use change processes have operated from the past until the present, and how these have affected the flow regime of the Ubaye River in the Barcelonnette area, located in the Alpes de Haute Provence, Southeast of France. This presented research is an original and novel contribution, as it addresses the diverse topics of flood hazard and climate and land use change. The coherent structure and logical scientific arguments clearly expressed throughout the book are enhanced by the sound and thorough literature review on these topics. The research objectives are distinctively addressed and followed throughout the text, and are supported by a thorough review of the available literature on this topic. The conceptual and methodological approaches are described in great detail and thus, the study is clearly reproducible. This research has been carried out in a specific catchment. However, the results are transferable and might be of highest interest also for other flood-related studies investigating climate and land use change pattern and their effect on the natural and human environment. Therefore I am confident, that this book is a rich source for other related studies.

Vienna, Austria, July 2012 Prof. Dr. Thomas Glade

Preface

This book aims to serve as an effective source of information, as well as a valuable basis for setting priorities and making further technical, financial, and political decisions regarding flood hazard assessment. In general, the main aim of this research was to develop and apply methods to identify and evaluate the response of flood events to land use and climate change. Flood hazard assessment was analyzed by computing a set of crisis scenarios (e.g. dyke failure), which are important information for civil protection purposes. Subsets of these scenarios were analyzed with distinct return periods as inputs into the hydraulic (1D) and hydrologic (1D–2D) models. Moreover, two severe flood events' characteristics (1957 and 2008) occurred in the study area were compared, and then flood propagation using hydrodynamic models (one dimensional (1D) HEC-RAS model and one-two dimensional (1D–2D) SOBEK model) were estimated, with respect to a variety of cross sections, river morphology, and different hydrographs. Additionally, this research gives a comparative overview of the major challenges faced with, like climate change and land use change, when dealing with flood hazards. Hydrodynamic simulation, together with trend detection climate change focusing on the analysis of climatological and hydrological parameters in a time series coupled with land use change estimation, led to the conception of a modern structure for flood hazard assessment. This research aimed to establish such a system, using a case study catchment to test and demonstrate its ability to perform under the typical conditions of mountainous catchments in France. The results of this research represent important fundamental information for policymaking, decision support, and flood hazard planning in any region, but are indeed particularly relevant for the Barcelonnette area. The results of this research have been presented at several international conferences, seminars, and workshops. Additionally, some articles are also already prepared (as listed in this book) and ready to submit in international journals. This research has been one of the most significant academic challenges I have ever faced. The University of Vienna, especially Prof. Thomas Glade, provided an intriguing foundation for me to work on physical aspects of the earth, focusing on hydrodynamic simulation (1D–2D modeling), climate change, and land use change. I am delighted and honored to have been able to work with Prof. Thomas Glade! From a scientific perspective, he taught me tremendously and is an amazing scientist with very strong knowledge in different aspects of science. Apart from this, he is a wonderful team-project manager and I believe that I have gained an immense amount of valuable experience from him that will be useful in my future. "Thank you Thomas".

Vienna, Austria, July 2012 Azadeh Ramesh

Acknowledgments

It is a great pleasure to acknowledge those who contributed to the success of this research.

My heartfelt thanks and foremost gratitude is expressed to my supervisor, Professor Thomas Glade, who kept my confidence and spirit high throughout the research process. His professional yet friendly ways of communication have helped me successfully to complete my research. His understanding, encouragement, and sustained commitment to letting me think for myself gave me complete freedom to develop the research while ensuring that it continuously moved forward in a positive direction. I owe particular thanks to him for his guidance on the direction and technique of my work. In addition, I deeply appreciate his valuable comments and support, as well as the critical and thorough insights he offered while reviewing my work. "Thomas, I thank you from the bottom of my heart".

My profound gratitude goes to the University of Vienna that partly funded this research. I would also like to express my sincere thanks to all staff members of the University for their kind support, including student affairs and all technical departments.

Out of deep appreciation I would also like to acknowledge Professor Victor Jetten for his exceptionally valuable contributions and comments which furthered the success and completion of this research.

I am grateful to Professor Samimi (University of Vienna) and Dr. Vaze (The Commonwealth Scientific and Industrial Research Organization (CSIRO), Australia) for offering comments and suggestions on this research.

I would like to thank Dr. Jean Philippe Malet for his great support in data acquisition.

I sincerely express my gratitude and thanks to Dr. Dinand Alkema for his academic and personal support, encouragement and guidance, and also for providing the desperately needed research tips throughout the research period in ITC. "Dinand, your belief in my ability and performance was very motivating".

I am deeply indebted to Ir. Gabriel Parodi for sharing his wealth of knowledge with me, which in turn stimulated my thinking and enabled me to develop an understanding of the subject.

I would like to express my earnest gratitude and thankfulness to all staff members of the Twente University, the faculty of Geo Information Science and Earth Observation (ITC), The Netherlands, who directly or indirectly helped me in the successful completion of my research and provided me with such a wide scientific environment, in particular, Dr. Cees Van Westen, Dr. Nanette Kingma, and Dr. Tom Loran. I highly appreciate their contribution and endless support during my research in ITC. The work in ITC for 6 months allowed me to learn a lot.

My sincere thanks and gratitude goes to Ir. Martijn Roetgerink for his continuous technical support throughout my research in ITC. He always was available to help me regarding to problems with ITC related IT especial SOBEK installation.

I express my appreciations to Deltares Software Centre (Delft Hydraulics), especially Ms. Sonny de Wit, for their useful consultation and technical support during my work in SOBEK simulation modeling.

A special thanks goes to ENGAGE (Geomorphological Systems and Risk Research's group) for their unending friendliness and always making a pleasant atmosphere to ensure that I never felt that I am alone in this process!

Last but not least, I owe my loving thanks to Alireza for the very special person he is. His endless love, smile, patience, understanding, inspiration, moral support, and persistent confidence in me never ceased, while putting up with my erratic work schedule, he still always managed to make life fun.

This research was carried out indirectly in the framework of the European "Mountain Risk" Project (http://www.unicaen.fr/mountainrisks). Furthermore, this research was financed by research grant (Forschungsstipendium der universität wien) provided by the University of Vienna.

Vienna, 2011 Azadeh Ramesh

Contents

Chapter 1
Rational

1.1 Introduction

This chapter intends to provide a better understanding of the problem statement regarding flood events on a global and local scale. Then, the hypothesis and objectives of this research are posed based on methodologies applied in this research. Finally, the work plan and structure of the thesis are presented and showed as a flowchart.

1.2 The Problem Statement

Scientific, political, and public opinion is agreed that we live in a period of accelerated flood hazards. Flood events as a destructive abundance of water have been a major concern for people populating the vicinity of rivers and water bodies since prehistoric times. Despite the fascinating developments in many areas of science and technology during the last decades, the hazard of flooding has not been eradicated. Flood events are not only one of the most widespread natural disasters which regularly cause large numbers of casualties, with rising economic loss, extensive homelessness, and disaster-induced disease, but they also are the most frequent disaster type (Fig. 1.1).

Damage caused by flooding is a rapidly increasing trend both globally and across Europe. In reference to the database reported by the Centre for Research on the Epidemiology of Disasters (CRED)—International Disaster Database (EM-DAT)—flood events have affected the highest number of people worldwide during the period 1900–2009.

According to this research center [1], flood inundations affected ~875 million people globally between 2001 and 2010 (Table 1.1), and estimated damages were ~US$142 billion (houses, infrastructure, crop, and livestock, etc.).

A. Ramesh, *Response of Flood Events to Land Use and Climate Change*,
Springer Theses, DOI: 10.1007/978-94-007-5527-7_1,

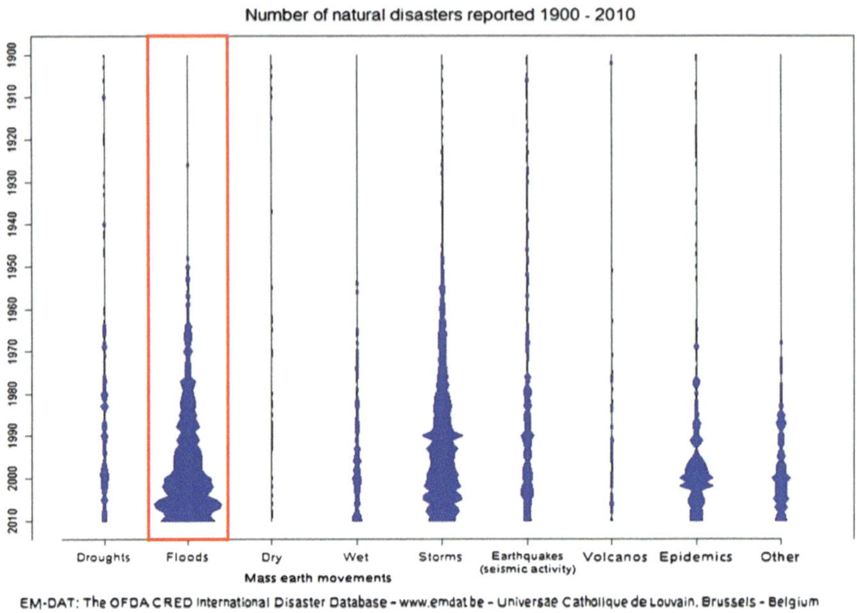

Fig. 1.1 Number of natural disasters reported 1990–2010 [1]

Table 1.1 The number of affected people and economic losses [2]

	No. of events	No. of affected people	Damage (in US$ 1000)
Period	World	World	World
1900–1980	581	407,380,734	2,496,946
1981–1990	546	469,855,416	46,648,071
1991–2000	966	1,463,420,671	233,075,277
2001–2010	1592	874,495,941	141,949,993
	Europe	Europe	Europe
1900–1980	53	594,454	658,250
1981–1990	51	839,522	10,482,300
1991–2000	147	4,679,707	41,674,964
2001–2010	209	3,205,507	41,793,882

In comparison to the previous decade (1991–2000), even if the number of affected people and economic loss both decreased by about 40 % during 2001–2010, the number of flood events would have still increased by 65 %.

Within the 20 year period from 1991 to 2010, the number of flood events more than doubled, the number of affected people nearly tripled, and the damage multiplied by a factor of more than seven in comparison to the 91 year period from 1900 to 1990 (Table 1.1).

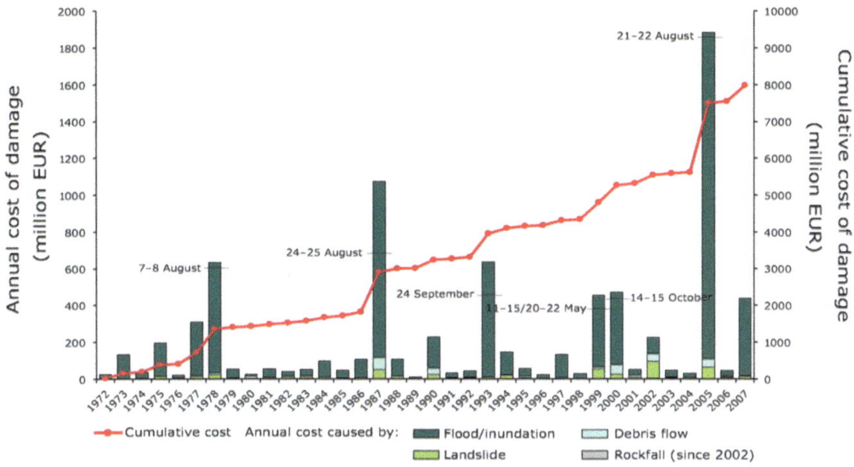

Fig. 1.2 Annual and cumulative cost of damage caused by different hazards [3–5]

From 1900 to 1990, ~9.2 % of worldwide flood events occurred in Europe. This proportion significantly increased between 1991 and 2000 (~15.2 %) and 2001–2010 (~13.1 %) [1, 2]. Figure 1.2 shows the annual and cumulative cost of damage caused by floods/inundations, debris flows, landslides, and rock falls from 1972 to 2007, as well as the total costs of the six major flood events around the world (Fig. 1.2), indicated by short horizontal lines and dates, which were reported by the Intergovernmental Panel on Climate Change [3–5].

During the last 30 years, 2,159 flood events were reported by the EM-DAT project, resulting in deaths of 206,303 people. Furthermore, flooding causes extensive damage to infrastructure and crops. The affected area is usually immense, but this depends on topographic features [6].

During the last three decades (1973–2002), the reported number of disasters caused by floods and registered in the EM-DAT database, which remained more or less stable during the first two decades, increased progressively during the last decade and experienced an exponential growth in the last three years.

In Western Europe, small disastrous floods are more frequently reported for Belgium, France, and Switzerland. From 1973 to 2002, the most frequently affected country by flood events was France, which accounts for almost 50 % of disastrous floods reported. Half of these flood events occurred in the last four years. Figure 1.3 shows the losses from major flood disasters in Europe between 1970 and 2008 [3–5].

France accounted for the greatest number of deaths for each decade and for 68 % of the total of reported deaths (between 1973 and 2002). Number of deaths over ten is reported in 41 % of the total number of disastrous flood events with reported deaths. The total number of deaths multiplied almost by three between the first and the

Fig. 1.3 Losses from major flood disasters in Europe between 1970 and 2008 (IPCC [3–5]

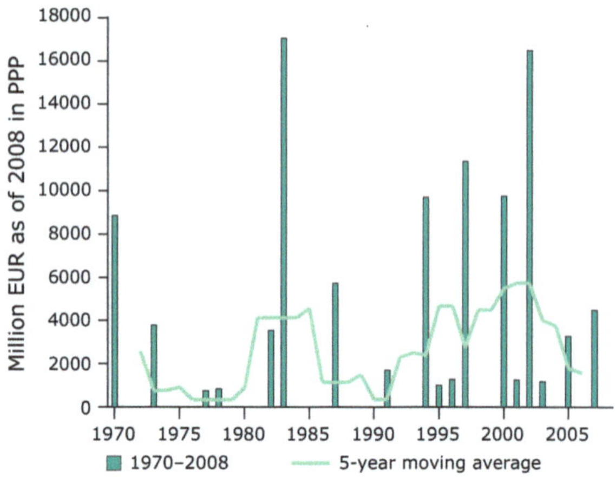

second decades (between 1973 and 2002), and remained stable during the third. In France, 64 % of the floods we have information on, report less than 1,000 people affected, and 21 % have between 1,000 and 10,000 affected. Bigger flood events have been reported in 1977 (25,000 affected), 1988 (more than 50,000 affected), and especially in 1995 (120,000 affected) and 1999 (200,000 affected) [7].

At the end of the nineteenth century, the Ubaye Valley was considered as the most dangerous torrential flooding area in France. The high frequency of debris flow interfered with the development of human activities. The correction of the active torrents was an emblematic challenge for the RTM services (Service de Restauration des Terrains de Montage—Mountain and Restoration Service) who were in charge of the torrential mitigation. At that time, torrentiality in the South French Alps resulted from intensive deforestation and land degradation [8].

According to the French Forest Office, more than 550 events have occurred since 1850. A total of 450 events were recorded as torrential events and 100 events were recorded as debris flows (Fig. 1.4).

Flageollet [9] indicated that there were six major inundations and 15 other hydrologic hazards such as debris flows, and overflows since 1950, though the numbers of references were around 428. Of this record, 13 % included debris flows, 12 % inundations, 2 % gullying, and 73 % were only torrential flows. Among these statistical reports, 15 events can actually be accredited as floods [9].

This area constitutes an elongated form. Elongated or linear villages are usually found along channels, rivers, or road sites that promote attenuated settlement forms. The elongated structure makes it highly dependent on structural measures such as dykes and levees to protect it against flooding.

The Ubaye Valley is representative of climatic, lithologic, geomorphologic, and land use conditions observed in the South French Alps, and is highly affected by flood hazards [10]. The city is mainly a tourist center hosting many ski lodges. The

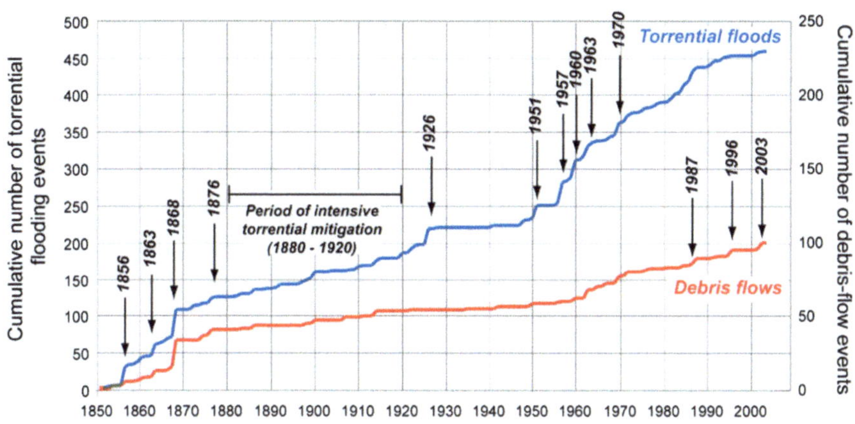

Fig. 1.4 Cumulative curves for floods and debris flows for the period 1850–2006 [9]

whole Barcelonnette Basin is known for its numerous slope (the slope gradients range from 10° to 20°) instabilities and other geomorphologic hazards [9, 11] and also for the very active flood hazards in this basin directly and indirectly affected by land use.

Unfortunately, a flood event as a natural phenomenon cannot be prevented. Some human activities such as increasing human settlements in floodplains, the reduction of natural water retention by land use, and climate change contribute to the increased likelihood of flood events [12].

1.3 Research Hypothesis and Objectives

In general, the main aim of the dissertation was to develop and apply methods to identify and analyze the flood hazard by computing of a set of crisis scenarios (e.g. dyke failure). The selection of a subset of these scenarios with distinct return periods as input into the hydraulic one-dimensional (1D) and hydrologic one–two dimensional (1D–2D) models was also estimated. Additionally, this research gives a comparative overview of the major challenges faced when dealing with flood hazards. It provides a methodology to investigate the probable climate change and land use change from the past until the present in the Barcelonnette area. Based on available data, and in accordance with the main aim, this work is supported by five specific objectives which contribute to investigate the posed research hypotheses as a pre-research prediction:

Hypothesis (a):	Interpolation techniques are suitable tools to control the quality of digital elevation model
Objective (a):	To find which interpolation method gives the best result based on statistical methods and watershed and stream network delineation
Hypothesis (b):	For any flood hazard assessment, simulation of variability of flood crisis scenarios and flood frequency analysis is necessary for civil protection purposes
Objective (b):	To apply different DEM resolutions, to investigate how much the accuracy of the results is influenced by provided input data
	To investigate river changes over time and see how river morphology affects the flow propagation
	To simulate variability of dyke scenarios and to estimate the capacity of the channel and dykes' resistance to the different volume of water. Additionally, to determine the hydraulic characteristics (i.e., propagation rate, flow depth, and velocity) for each simulation
	To model different hydrodynamic models and to investigate advantages and disadvantages for any future research in similar characteristic river channel and flood plain
	To analyze probability distribution and return period analysis
Hypothesis (c):	Topographic elements can be used as strategic flood defense systems especially to delay flood waters in order to evacuate people
Objective (c):	To simulate the variability of flood crisis scenarios for dyke failure
Hypothesis (d):	Climatic variables influence flow distribution
Objective (d):	To find any probable abrupt points and trend analysis in spatial distribution of observed data sets (climatologic and hydrologic parameters), statistical methods should be implemented
Hypothesis (e):	Flow propagation in the Ubaye river is affected by anthropogenic activities and land use change
Objective (e):	To detect land use change from the past until present. GIS process analysis should be used to ensure any affects on flow regime of the Ubaye river

1.4 Work Plan and Structure of the Thesis

This thesis is structured into eight chapters. Chapter 1 provides a foundation for the research and introduces the problem and subsequently, the hypothesis, the aim, and objectives of the research, the workflow and structure of the thesis. Chapter 2 gives a comparative overview of the major challenges faced when dealing with flood hazards. Chapter 3 is constituted by the conceptual and methodological approaches leading the research dealing with digital elevation model (DEM) quality control modeling, interpolation technique comparisons, and hydrodynamic simulation with respect to different flood crisis scenarios, frequency analysis, climate change detection, and land use change estimation from the past to the present. Chapter 4 describes the study area and the Ubaye River characteristics. Chapter 5 covers available data. Chapter 6 explores the results of SOBEK and HEC-RAS simulations, climate change detection, and statistical analysis for land use change and attempts to estimate the feedback and interaction between all achieved results scientifically. Chapter 7 provides an overall conclusion derived from the results of the applied methodology and assessments performed. Additionally, key recommendations are discussed. Chapter 8 highlights some directions and perspectives for future investigation for the Barcelonnette area.

Hydrodynamic simulation, together with trend detection in climatologic and hydrologic parameters and land use change estimation, led to the conception of a modern structure for flood hazard assessment. This thesis aims to establish such a system, using a case study catchment to test and demonstrate its ability to perform under the typical conditions of mountainous catchments in France.

In this research, the focus was on flood hazard assessment based on hydrologic, land use, and climatic parameter analysis in a time series. Within hydraulic modelling, the output is affected to a large extent by model input parameters. In this mention, first the testing of the quality of the DME using interpolation techniques was important in order to accomplish a successful modeling. Then, all geometric data were prepared in the HEC-GeoRAS model.

Before starting hydrodynamic modeling, based on available data, land use maps from 1957 to 2008 were analyzed to gain better insight into river morphology change over time and to access a variety of roughness derived from different land use maps to assign to cross-sections in flood modeling. Then, the focus was on the preparation of hazard maps and frequency analysis. The main purpose of this part was to compare some severe flood events that occurred in this area, and then to estimate flood propagation using hydrodynamic models (1D HEC-RAS model and 1D–2D SOBEK model), with respect to a variety of cross-sections, river morphology, and different hydrographs. Moreover, the preparation of different hydrologic scenarios for civil protection purposes was considered in this study area: (1) In the case of the DEM scenario, the model was run with different DEM resolutions to understand how the grid resolution of a DEM affects the flood characteristics when simulated by a 1D and 1D–2D flood models. (2) In the case of the river change scenario, the aim was to evaluate river change scenarios in order to know how river channel changes affect the flood, using different profiles and different discharge values. (3) In the case of the dyke failure scenario, the major aspect was realizing the possible effects of dyke failure. (4) Finally, the main aim in the model scenario, was to compare two different hydrologic models, including HEC-RAS (1D) and SOBEK (1D–2D), in order to know what the advantages (or strengths) and disadvantages (or weaknesses) of these models are. In the last part of the hydrodynamic simulation, the probability distribution and the return period were analyzed.

Climatic parameters are one of the most important triggering factors for a flood event. The trend analysis of hydrologic and meteorologic series is important, and even more relevant when considering the regional effects of global climate change. Therefore, in the framework of climatic parameter analysis, any probable abrupt points and trend analysis in the time series of climatologic and hydrologic parameters using statistical methods were explored. Based on available aerial photos, land use change was detected and statistically analyzed from 1957 to 2008 using GIS techniques. Finally, to gain feedback and understand the relationship among climate change, land use change, and flood events, all achieved results were reviewed. Therefore, the thesis is divided into three main phases:

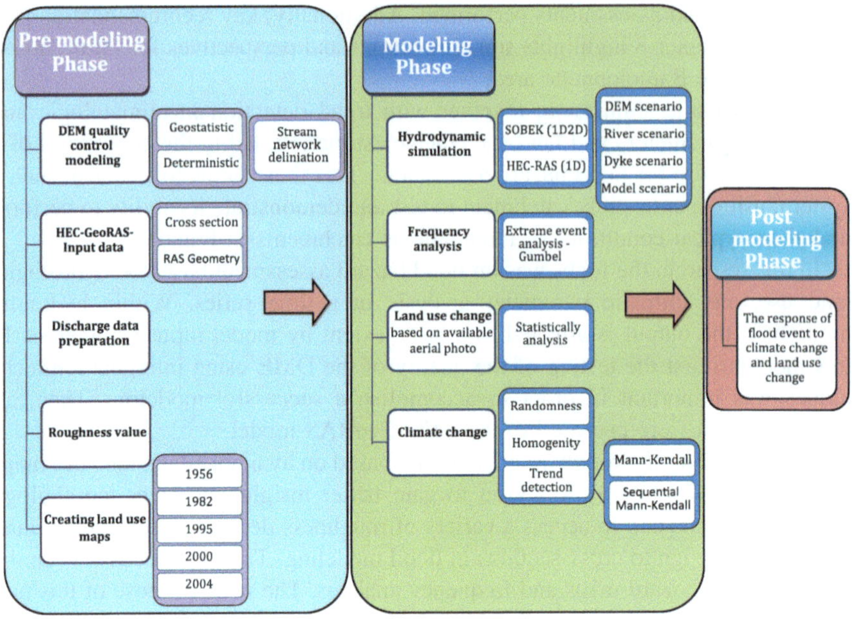

Fig. 1.5 Principle methodology flowchart

Premodeling phase: The tasks in this part included: DEM Quality control modeling using deterministic and geostatistical methods, preparation of RAS-geometric layers using HEC-GeoRAS model, preparation and modeling of discharge database of observed data, and preparation of roughness values derived from different land use maps.

Modeling phase: The main task in this part included: flood hazard assessment using hydrodynamic modeling (1D and 1D–2D), climatologic and hydrologic parameters analysis and trend detection using statistical techniques, land use change analysis using GIS techniques and statistical methods.

Postmodeling phase: The aim of this part was to study the interaction and feedback among land use change, climate change, and flood events. The principal methodologic steps performed during this research are shown in Fig. 1.5.

1.5 Chapter Summary

This chapter has laid the foundations for the research and introduced the problem, hypothesis, aim, and objectives of the research, and lastly, the workflow and structure of the thesis. As a problem statement, it discussed how flood events have challenged people's lives throughout the world. Statistically, it has estimated the number of affected people as well as economic losses such as damage to houses, infrastructure, crops, etc., both for the world and Europe. Five research hypotheses as pre-research predictions were proposed in this chapter. To respond to the problem statement and the hypothesis behind this study, the research put forward five objectives: first, to increase the quality of the DEM; second, to detect flood hazard assessment; third, to simulate the variability of flood crisis scenarios for dyke failure; fourth, to find any probable abrupt points and trend analysis in spatial distribution of the observed data set; and finally, to detect land use change over time. The research approach was then designed and described based on five objectives in three phases of premodeling, modeling, and postmodeling. The research was formulated and the thesis structure outlined in eight chapters.

The next chapter provides a theoretical background to the DEM, interpolation techniques, flood hazard, climate change, and land use change.

References

1. CRED/EM-DAT data, Centre for research on the epidemiology of disasters (CRED)—international disaster database (EM-DAT), Available online at: http://www.emdat.be/database (Verified 28 Nov 2011)
2. Martinis S (2010) Automatic near real-time flood detection in high resolution X-band synthetic aperture radar satellite data using context-based classification on irregular graphs, Dissertation der Fakultät für Geowissenschaften der Ludwig-Maximilians-Universität München
3. Intergovernmental panel on climate change (IPCC) (2007a) Climate change 2007: the physical science basis. Contribution of working group I to the fourth assessment, report of the intergovernmental panel on climate change, Cambridge University Press, Cambridge and New York, p 996
4. Intergovernmental panel on climate change (IPCC) (2007b) Summary for policymakers.In: Parry ML, Canziani OF, Palutikof JP, van der Linden PJ, Hanson, CE (eds) Climate change 2007: impacts, adaptation and vulnerability, contribution of working group II to the fourth assessment report of the intergovernmental panel on climate change, Cambridge University Press, Cambridge pp 7–22
5. Intergovernmental panel on climate change (IPCC) (2007c) Climate Change 2007: synthesis report, contribution of working group to the fourth assessment report of the intergovernmental panel on climate change (IPCC). Cambridge University Press, Cambridge
6. Guha-Sapir D (2006) Climate change and human dimension: health impacts of floods, international workshop on climate change impacts on the water cycle, resources and quality 25–26 Sept 2006, Brussels
7. Hoyois P, Guha-Sapir D (2003) Three decades of flood in Europe: a preliminary analysis of EMDAT data, WHO collaborating centre for research on the epidemiology of disasters (CRED), Catholique University of Louvain
8. The french observatory—Observatoire Multidisciplinaire des Instabilités de Versants (OMIV), Data Access—Barcelonnette area. http://eost.u-strasbg.fr/omiv/data_access_Barcelonnette
9. Flageollet J-C (1996) The time dimension in the study of mass movements. Geomorphology 15:185–190
10. Flageollet J-C, Maquaire O, Martin B, Weber D (1999) Landslides and climatic conditions in the Barcelonnette and Vars basins (Southern French Alps, France). Geomorphology 30:65–78

11. Weber D (1994) Research into earth movements in the barcelonnette basin. In: Casale R Fantechi R, Flageollet JC (eds) Temporal occurrence and forecasting of landslides in the European community, final report., contract EPOCH, European Commission, vol 1 pp 321–336
12. Directive 2007/60/EC of the European parliament and of the council of (23 Oct 2007) on the assessment and management of flood risks text with EEA relevance, Official Journal of the European Union, L 288:0027–0034

Chapter 2
Theoretical Background

2.1 Introduction

With respect to the literatures and previous researches, this chapter gives a comparative overview of the major challenges faced when dealing with flood hazard. First, definition, source and implication of digital elevation model are explained, and then the focus is on application of different interpolation techniques in drainage network estimation, coupled with advantages and disadvantages of these methods and their application in different research fields. Flood hazard, concepts, definition, types, and causes are defined in the next parts. This chapter also provides a background to climate change and land use change from global to local (Barcelonnette area) which was estimated and interpreted based on different scientific researches.

2.2 Digital Elevation Model: Sources and Implications

Topography representing as digital elevation model (DEM) is an important land-surface characteristic that affects most aspects of the water balance in a catchment, including the generation of surface and sub-surface runoff; the flow paths followed by water as it moved down and through hill slopes and the rate of water movement. All of the spatially explicit fully distribute hydraulic and hydrological models as well as hydrological decision support systems use topography (represented by DEM of the area modeled) to derive bathymetry [1].

Many applications depend on the shape represented by the DEM rather than the absolute elevation values. These include hydrological applications, for which an accurate representation of surface drainage structure is critical [2, 3]. Computation of terrain corrections to observed gravity data and remotely sensed data is another important application that depends primarily on the representation of terrain shape.

A. Ramesh, *Response of Flood Events to Land Use and Climate Change*,
Springer Theses, DOI: 10.1007/978-94-007-5527-7_2,

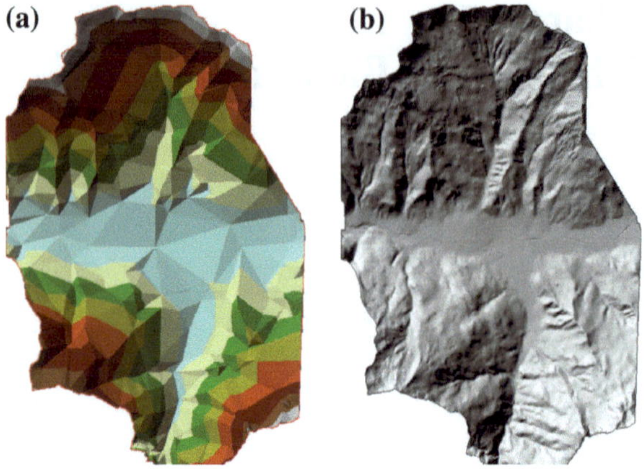

Fig. 2.1 Comparison between two terrains. **a** Irregular DEM. **b** Gridded DEM

The measures of shape most commonly used are slope and aspect of the down slope direction [3].

According to Vaze et al. [1] and Teng et al. [4] DEM is used to derive some key information which is important in distributed hydraulic and hydrological models; such as flow paths; dispersion; and accumulation of water, terrain slope, drainage networks, drainage divides, and catchment boundaries. Vaze et al. [1] also compared the traditional methods of topographic maps, field surveys or photographic interpretation with the application of DEMs. Vaze et al. [1] concluded that DEM is an effective way to present ground surface and extract the hydrological features; thus bringing advantages in terms of processing efficiency, cost effectiveness, and accuracy assessment. Burrough and McDonnell [5] defined DEM as "any digital representation of the continuous variation of relief over space". Wood [6] also defined DEM as "a computer representation of the earth's surface which is provided a base data set from those topographic parameters that can be digitally generated". Alternatively, the DEM could be used in mapping the possible locations of an endangered species whose habitat is altitude dependent. Here, the resultant value for each grid cell may be either the elevation of the center point of the grid cell or the average height of the area covered by the grid cell. The actual characteristics of a grid can be adapted to suit its major application. For instance, the application of the grid may require that all local high points (hills or mountains) in the source data be retained in the grid. Technically, Olivera et al. [7] have another definition for DEM; they defined DEM as an array of squared pixels or cells with an elevation value specific to each pixel. They said that DEM is commonly used in automated drainage analysis methodologies due to its inherent simplicity characteristics of the data structure. Point elevation information in DEM which is a representation of the terrain, can be of two types—*a. irregular and b. gridded* (Fig. 2.1) [3]; a. An *irregular* spaced DEM is often interpreted as a triangular irregular network (TIN). TIN represents a surface as a set of irregularly located points linked to form a network of triangles with z-values stored at the nodes.

The terrain is simulated as a series of planar triangular facets produced by joining all the adjacent points. The assumption that the surfaces are planar is adequate if the points have been chosen at changes in grade in the terrain. The accuracy of the TIN model can be improved by the addition of break lines. These lines represent discontinuities in the terrain surface such as cliffs, ridges, and streams and they indicate where interpolation between adjacent points is invalid [3]. b. In a *gridded* DEM, the elevation points are spaced at a regular interval to create a grid or lattice [3]. DEM represents a surface as a regular grid of locations with sampled or interpolated values.

By incorporating stream lines and cliff lines, a gridded DEM can represent all the discontinuities that can be represented by a TIN and has the significant advantage of being directly compatible with other sources of natural resource data in grid form. It is also readily used by many grid-based applications [3].

Where the gridded DEM has been derived from a primary data source such as contours or a TIN, then the direct relationship between the elevation value and the actual value on the ground, at the respective location, is dependent on the algorithm used to interpolate the grid and the resolution of the grid itself.

The DEM, as its name indicates, is a model of the elevation surface. However, the DEM is often not treated as a model, but is commonly accepted as a "correct" representation of the earth's surface. But, it should not be neglected that DEM data, like other spatial data sets are subject to error. The estimation of errors in a DEM is often not evaluated by DEM users and applicants [8]. Where the DEM is directly observed from aerial photogrammetry or a field survey, the elevation value is truly representative of the value that is found on the ground at the location of that point, provided there is no significant measurement error.

Significant measurement errors can arise due to ground cover by vegetation and buildings, especially when elevations are measured by aerial and space-borne platforms. These measurements are also affected by complex terrain [9]. This is necessary if the DEM were to be used for aircraft flight planning where the minimum flying height of the aircraft is critical. This could also aid the sitting of signal transmitters and receivers, although high points could then have errors in horizontal position by up to half of one grid interval [3]. In areas away from peaks, these values are approximately the same. In general, Burrough [10] and Wise [23] listed the possible errors in DEM data sets as the following:

- Data errors due to the age of data, incomplete density of observations, or results of spatial sampling.
- Measurement errors such as positional inaccuracy, data entry faults.
- Processing errors such as numerical errors in the computer, interpolation errors, or classification and generalization problems.

DEM is usually produced form sampled or observed data points that are used as its source. The contour lines themselves may represent a model of terrain. It can also be acquired directly, for instance photo grammatically from a stereo model or indirectly, from analog cartographic data, satellite images or by field surveying, and so on. Ideally, the data sources are applied without application of interpolation techniques. Application of interpolation techniques is not necessary if the data sources are very accurate, precise, and have high density, especially if the data are derived directly into a regular grid (DEM) [3, 11].

The quality of a derived DEM could vary largely depending on the data source and the interpolation technique. The desired quality also depends on which application the DEM is used; although, a DEM created for one application is often used for other purposes. Therefore, to create any DEM, it is necessary to consider the best available data sources and the best processing technique.

Spurious sinks or local depressions in DEMs are frequently encountered and are a significant source of problems in hydrological applications. Sinks may be caused by incorrect or insufficient data, or by interpolation techniques that does not enforce surface drainage. They are easily detected by comparing elevations with surrounding neighbors [11]. Therefore, detection of spurious sinks features or local depressions in DEMs can lead to improvements in DEM generation techniques as well as detection of errors in source data as indicated above [12].

More subtle drainage artifacts in a DEM can be detected by performing a full drainage analysis to derive catchment boundaries and streamline networks, using the drainage networks creation techniques [13]. Since, applications of DEMs depend on representations of surface shape and drainage structure, absolute measures of elevation error do not provide a complete assessment of DEM quality [2]. Based on the above description about DEM error, it is necessary to make it clear that, in science, the word "error" does not carry out the usual meaning of the term "mistake" or "blunder". The error in scientific measurements means the inevitable uncertainty. As such, errors are not mistakes; the scientists cannot eliminate them; therefore, the best thing to do is ensure that errors are as small as reasonably possible with reliable estimate of how large they are [14]. To this purpose, the starting point in this research was statistical, spatial and hydrological controlling, estimating, and correcting the possible errors in the DEM [15] which will be explained in the next sectors.

2.3 Interpolation Techniques in Drainage Network Estimation

Interpolation methods often assume data points are correct and accurate, but it is assumed that these data points may be subjected to error [5]. The models application to estimate the unknown points may predict the data points exactly (go precisely through the sample data points) or inexactly (approximate the values at the data points). If the observed data points are relatively sparse and irregular or widely spread, interpolation needs to be more sophisticated than for dense, regularly spaced data. The principles of interpolation are shown in Fig. 2.2.

However, regularly spaced data may be subject to bias due to intrinsic frequencies in the data [5]. An interpolation method is working globally, if all data points are evaluated in the interpolation. Local interpolation techniques use only

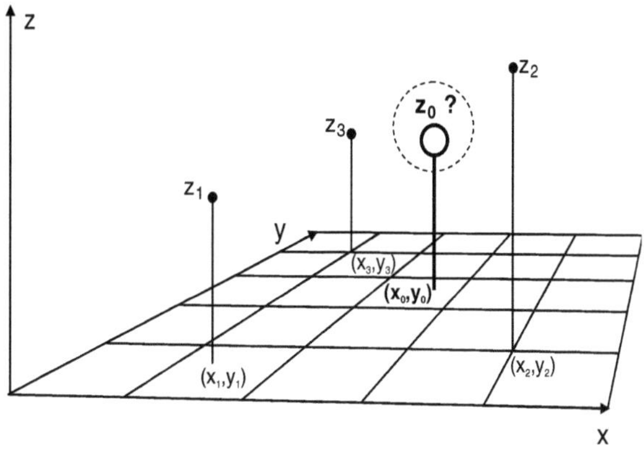

Fig. 2.2 The principles of interpolation technique [16]

data points in a certain neighborhood of the estimated point [16]. The selection of interpolation methods depends primarily on the nature of the variable and its spatial variation [16].

There have been many studies that compare the effectiveness of alternative interpolation techniques, using a wide range of different test datasets and conditions. Overall, it has been found that a number of well-defined factors have a major influence on the quality of interpolation: data measurement accuracy; data density; data distribution; and spatial variability.

These factors are fairly predictable findings, but prior examination of each of these elements may assist in choosing the most appropriate technique for the problem at hand and/or be used in guiding sampling of new or supplementary data sets. Interpolation quality can often be substantially improved through the use of ancillary information, such as remote sensing data or additional environmental information (e.g., location of stream networks) [17]. Having obtained the best possible data set, within budget and time constrains, achieving the maximum usage and value is very important. Hence more general spatial interpolation is required [17]:

- To convert from one level of data resolution or orientation to another (resampling). Usually, resolution is reduced to the coarsest in a set, but resolution can be increased using a suitable interpolator.
- To convert from one representation of a continuous surface to another, e.g.,TIN to grid or point or contour to grid [17].

There are two interpolation techniques; Deterministic and Geostatistical:

- Deterministic interpolation is directly based on the nearby measured values or on specified mathematical formulas that determine the smoothness of the resulting surface.
- Geostatistical interpolation is based on statistical models that include autocorrelation (statistical and spatial relationships among the measured points).

Different interpolation techniques characteristics are summarized in Table 2.1 [17, 19]:

Table 2.1 Different interpolations characteristics (modified information based on [18, 19])

Method	Principle	Advantages	Disadvantages	Best suited scenario	Schematic
Nearest neighbor (NN) and thiessen polygon	Selection of values at closest data point	Ease of use	Inaccurate in less densely sampled scenarios	Densely sampled environmental data	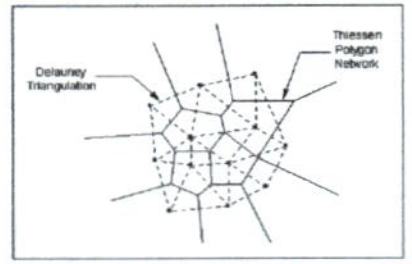
Triangulated irregular network (TIN)	Set of conterminous with a mass factor is used to define the space	Ability to describe the surface at different levels of resolution	In most cases, required visual inspection and manual control of the network	Dense and moderate distribution of data points	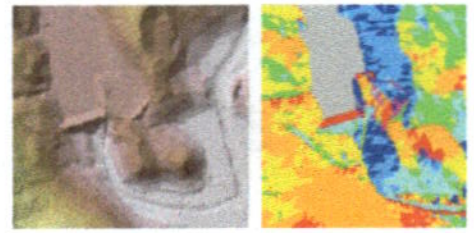
Polynomial regression (PR)	Fits the variable of interest to the linear combination of regressor variable	Simple model	Model has poor ability to predict outside the range of data points	Moderately dense sampling with regard to global variation	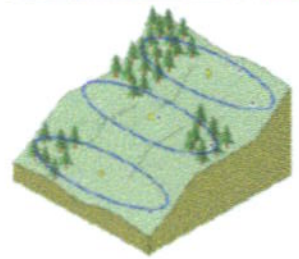
Global polynominal interpolation (GPI)	Works by capturing coarse-scale patterns in the data, and fitting a polynomial	Computationally less intensive	Estimation errors increase exponentially with increasing complexity	Regions having sparse data points and simple data patterns	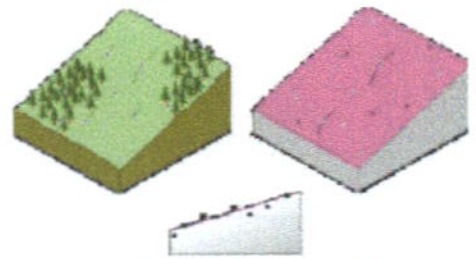
Local polynominal interpolation	Similar to GPI, but the curve is fitted to a local subset defined by windows	Can interpolates short range variations	Misses the global trends in data	Well-distributed data with no discontinues	
Trend surface analysis (TSA)	Separates the data into regional trends and local variations	Assists in removal of broader trends prior to further analysis	Edge effects and multi-co linearity caused by spatial autocorrelation	Important local trends and not so important global trends	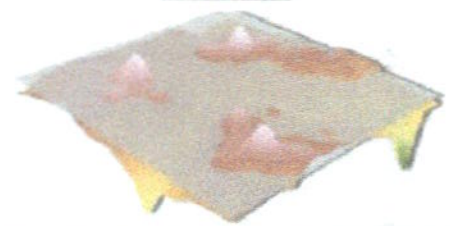

(continued)

Table 2.1 (continued)

Method	Principle	Advantages	Disadvantages	Best suited scenario	Schematic
Inverse distance weighting (IDW)	Linear combination of known points, weighted inversely by distance	Ease to use, and works well with noisy data	Spatial arrangement of samples does not affect weights	Moderately dense sampling with regard to local variation	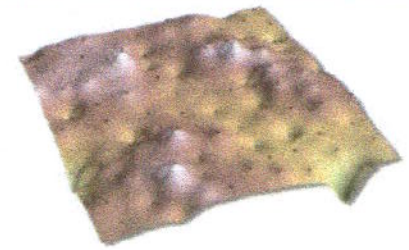
Splines	Fits a smooth curve to a series of data points	Visually appealing curves or contour lines	May mask uncertainty present in the data	Irregularly spaced data	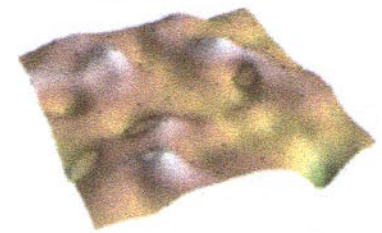
Radial basis functions (RBF)	Similar to the principle of splines, except the curve is not smooth	Required fewer samples	Required good coverage of input space, and not suited for extrapolation	Regions with well distributed data points, though sparse	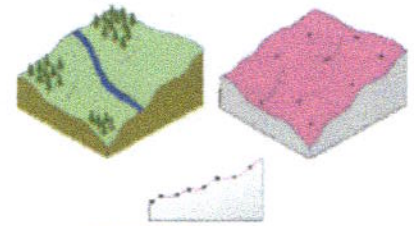
Artificial neural networks (ANN)	Learn complex patterns in the available data using Sigmoid functions etc.	Ability to learn and generalize data; works well with sparse data distributions	Risk of poor interpolation caused by over-learning or under- learning	Regions ranging from sparse irregularly distributed data to well distributed data	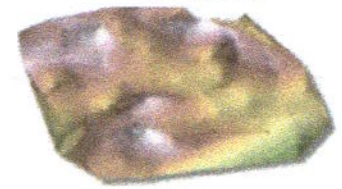
Kriging	Similar to the principle of IDW; however, additionally accounts for the spatial arrangement	Best linear unbiased spatial predictor: and no edge effects resulting from trying to force a polynomial to fit the data	Sophisticated programming required; and problems of non stationary in real world data set	Well-distributed data with no discontinues	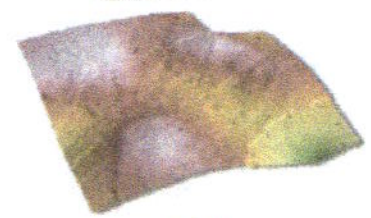
Topo to raster	Not specified	Based on iterative finite difference methods. Interpolates a hydrologically "correct" grid from a set of point, line and polygon data	Requires contour vector data as input. Available in ArcGIS based on Hutchinson's ANUDEM program	well-distributed data points	

Application requirements play an important role to expected characteristics of the used DEM. For example, it is not necessary to use high geomorphologic quality of DEM for regional, small-scale analysis and for calculating average altitudes. But geomorphologic accuracy is more sensitive for visibility analyses and even more for analyses that uses algorithms bases on derivates like slope, aspect, cost surface, drainage, path simulation, and so on [15].

Interpolation techniques based on the principles of spatial autocorrelation, which assumes that objects close together are more similar than objects far apart. On the edges of the interpolated area extrapolation is also reasonable. Unfortunately, no one of the interpolation techniques is universal for all data sources, geomorphologic phenomenon or other purposes. It is necessary to be aware that in the praxis, different interpolation methods and interpolation parameters on the same data sources lead to different results. The best chosen algorithms on fair data sources should not differentiate much from nominal ground, that is idealization of our desired model and which is commonly similar to the actual Earth's surface. Divergences between results of interpolation and from nominal ground are especially consequences of the following circumstances [15]:

- Available data sources do not approximate terrain distribution, density, accuracy, etc., of the sources is not appropriate.
- Selected interpolation algorithm is labile (is not robust enough) on the employed data sources.
- Chosen, interpolation algorithms or data structure, are not suitable for selected terrain geomorphology or application.
- Perception or interpretation of the Earth's surface (better: nominal ground) is not the same when more DEM operators work on the same problem; operator's own imagination is common and a reasonable problem in DEM production [15].

Application requirements play an important role to expected characteristics of the used DEM. For example, it is not needed high geomorphologic quality of DEM for regional, small-scale analysis and for calculating average altitudes. But geomorphologic and hydraulic accuracy is more sensitive for visibility analyses and even more for analyses that uses algorithms bases on derivates like slope, aspect, cost surface, drainage, path simulation, bathymetry of the river, and so on [15].

In the most cases, a very high quality DEM should cover all application demands. So it is preferable to find a good and robust interpolation algorithm. Even if a more generalized surface is required, DEM with high detail can be simplified to the required quality. It should be noticed that appropriate generalization methods are very important for producing required DEM. Commonly these methods are complex [15].

The quality of a derived DEM can vary greatly depending on the data source and the interpolation technique. The desired quality depends on the application for which the DEM is to be used, but a DEM created for one application is often used for other purposes.

Any DEM should therefore be created with care, using the best available data sources and processing techniques. Efficient detection of spurious features in DEMs can lead to improvements in DEM generation techniques, as well as detection of errors in the source data, as indicated above.

Since most applications of DEMs depend on representations of surface shape and drainage structure, absolute measures of elevation error do not provide a complete assessment of DEM quality [20, 21]. A number of graphical techniques for assessing data quality have been developed. These are nonclassical measures of data quality that offer means of confirmatory data analysis without the use of an accurate reference DEM [23].

2.4 Flood hazard: Concepts and definitions

World meteorological organization (WMO) [22] reported that floods events are among the most common, costly and deadly of natural hazards and have been a major concern to people residing near rivers and coastal areas throughout history. Despite the great developments in science and technology in recent decades, the hazards of flooding have not been eradicated [24].

According to the European Parliament and the Council of the European Union's directives, the "floods" means the temporary covering by water of land not normally covered by water. This shall include floods from rivers, mountains torrents, Mediterranean ephemeral water course, and floods from the sea in the coastal areas and may exclude floods from sewerage systems [25].

UN-ISDR [26] also defined flood hazard as the "potentially damaging physical events, phenomenon or human activity that may cause the loss of life or injury, property damage, social and economic disruption or environmental degradation".

Wassef and Aysan [27] discussed the geographical distribution of disasters and compared the flood hazard to other natural phenomenon in their research and found floods as the worst phenomenon that affected people's lives. According to Davis [28] investigation, there were 118 major flood events from the biblical deluge to the present, and Wasseff [29] also listed 87 floods during the period of 1947–1991 which resulted in homelessness of at least 50,000 people. In 1887, one of the worst recorded flood event occurred along the Yellow River in China. This flood resulted in at least 1.5 million deaths and caused as many as 10 million homeless ([28]. During 1982–1991, flood events left approximately 21,000 deaths per year and affected 73 million persons per year [27]. The investigation results of Wijkman and Timberlake [30] and Wasseff [26] indicate that since 1960, each passing decade shows an increase in the number of flood disasters, and the number of people which were affected by these flood events. Yen and Yen [31] has also shown that relative flood damages in the USA as a fraction of the annual Gross National Product, which expressed a declining trend between 1929 and 1993. More recently, flooding has occurred in Pakistan in 2010, which has directly affected approximately 20 million people, mostly by destruction of property, livelihood and infrastructure, and left about 2,000 dead [22].

Climate change as huge natural variability and long-term persistence phenomenon is realized as an important factor which affects on flood events but make it difficult to realize any trends in extreme weather events. During recent decades, flood damages in Europe have increased considerably [32]. According to previous investigations which have focused in Europe, in a large part, this increasing trend in flood hazard can probably be also attributed to human behavior, such as developing urbanization in flood plains [33].

2.5 Types and Causes of Flood Events

Lowland flood-prone areas can be found all over the world, along the coasts, in the river floodplains, and as inland depressions. Generally, they are basically unsuitable for development by their nature; sensitive areas with high physical conditions and environmental value. However, due to their strategic location and/or suitability for agriculture activities, there is often a tremendous pressure to develop these areas for various type of land use beneficial purpose. Therefore, a rapid population growth, significant increase in agricultural exploitation, urbanization, and industrialization may be observed in these lowland and flood-prone areas. Due to this, Schultz [34] mentioned that such areas become increasingly vulnerable from extreme weather conditions that will have their effect on the requirements for any drainage and flood-hazard management.

Floods include river floods, urban floods, coastal floods, and sewer floods. They can be caused by meteorological major causes such as; (a) Intense and/or long lasting precipitation or rainstorms on a small scale which cause flash floods. According to WMO [35, 36], flash floods typically occur by convective precipitation at high intensity, in short duration (less than 2–6 h) and limited aerial extent (less than 1,000 km^2); (b) widespread storms, which are caused by flooding on a regional scale. For regional flooding, the range of rainfall duration may differ from several days to a week or, in exceptional cases (in very large watersheds), may be associated with multiple storms occurring over a period of several months, such as in the 1993 flood in the Upper Mississippi River basin or the 1998 flood in the Yangtze River basin. Other meteorological causes of floods may be grouped into (c) Snowmelt; and (d) Reduced conveyance due to ice jams or landslides broad categories. Snowmelt floods are the result of three factors. The existence of the snowpack (aerial extent and depth), its condition (temperature and water content) and the availability of energy for melting snow. Snowmelt occurs when energy is added to a snowpack at 0 °C. In snow-dominated regions, some of the largest floods are caused by warm rain falling onto a snowpack at this temperature. In very large, snow-dominated watersheds, the annual peak flow is nearly always caused by snowmelt, whereas either snowmelt or rainstorms can cause the annual peak in small or medium-sized watersheds. In the cold regions, extreme high water stage can be caused by snow obstructing very small channels or ice jams in large rivers. In the case of rainfall or snowmelt flooding, natural processes can be exacerbated by watershed changes that enhance runoff production, cause flows to move more rapidly into the channel, or cause flows to move more slowly or more quickly within the channel. Thus, deforestation, overgrazing, forest or bush fires, urbanization and obstruction, or modification of drainage channels can be extensive or so severe as to have a significant effect on flooding.

Tropical cyclones produce hazards from storm surges, to wind and river flooding. Earthquakes and volcanic eruptions can produce landslides that cause flooding by damming rivers. Volcanic eruptions are associated with hazardous mudflows, and volcanic ash may cause flooding by choking river channels. From a natural hazard perspective, there are important similarities between river flooding; lake flooding; flooding resulting from poor drainage in areas of low relief; and flooding caused by storm surges (storm-induced high tides), tsunamis, avalanches, landslides, and mudflows. All are hazards controlled, to some extent, by the local topography, and to varying degrees it is possible to determine hazard-prone locations. Mitigation and relief efforts are also similar [22].

Although the influence of these causes could be strongly affected by some other factors, for example, rainfall and sea levels are two major causes of floods, and are natural and uncontrollable phenomena. However, these natural phenomena which results in flood damage are very much influenced by human behavior such as: deforestation in the upper catchment area, straightening of rivers and suppression of natural flood plains, inadequate drainage practices, and extensive building in high risk flood areas [37]. Apart from nonclimatic factors that affect flood events, there have been adverse flood-hazard changes due to climate change, e.g., increasing potential for intense precipitation in the warming world. The scientists in WMO also believe that there is no doubt, with currently available scientific data, that climate is changing in the sense of global warming. Global warming, whatever be the eventual magnitude, will very certainly affect the location, frequency, and strength of meteorological hazards [22].

Nonetheless, this research deals with the Ubaye River floods, thus river flooding will be the focus of this research. Floods along the rivers are natural and an inevitable part of the resident's life. River Floods depend on precipitation intensity, volume, timing, antecedent conditions of rivers, and their drainage basins (e.g., presence of snow and ice, soil character, wetness, urbanization, and existence of dykes, dams, or reservoirs) [38–40]. Some floods occur seasonally with winter or spring rains; coupled with the melting of snow, the river basins would fill with too much water too quickly. Torrential rains from decaying hurricanes or tropical systems can also produce river flooding [41]. Communications from the commission to the council (the European parliament) report that river floods also may occur whenever the capacity of the natural or man-made drainage system (such as dams, dykes) is unable to cope with the volume of water generated by rainfall, or when flood defenses fail. These are of the most common causes for river flooding in Europe [37]. Damaging floods will increase magnitude, when the capacity of the main conveyance of the river channel is exceeded. The main conveyance may be the primary channel of a river without dykes or the area between the dykes for a river with dyke. The capacity of the channel may be exceeded as a result of excessive flow distribution or ice or debris jams which could block the flow.

According to theoretical background and previous research regarding to flood, it is widely acknowledged that anthropogenic behaviors affect floods and flood hazards. Land use change can affect the amount of runoff for a given storm and the rapidity with which it runs off. Human occupancy of floodplains increases their vulnerability due to exposure to flood hazards. Dams, levees, and other channel alterations affect flood characteristics. Increased occupancy of floodplains and larger floods due to deforestation and urbanization has been attributed also to increase flood damage. Deforestation and urbanization increase flooding because they decrease the capacity of the land to absorb rainfall. It is customary to assume that flood hazards are stationary, i.e., they do not change with time. Climate change, anthropogenic influences on watersheds or channels, and natural watershed or channel changes have the potential, however, to change river flood hazards.

2.6 Climate Change and its Variability in Europe

In the past, destructive flooding maybe caused by only extremely heavy precipitation. Nowadays, less extreme precipitation may lead to a serious floods event [42]. Therefore, apart from nonclimatic factors that have affected flooding, there have been adverse flood-hazard changes due to climate change, e.g., increasing potential for intense precipitation in the warming world. The scientists in WMO also believe that there is no doubt, with currently available scientific data, that climate is changing in the sense of global warming. Global warming, whatever be the eventual magnitude, will very certainly affect the location, frequency and strength of meteorological hazards [22]. Many investigators believe that climate change is expected to have substantial impacts on hydrology on global, regional, and local scale [43, 44]; Andréasson et al [43]; [38–40, 40, 46]. There is also strong evidence that rainfall changes associated to global warming are already taking place on global and regional scale. The trend was globally positive throughout the twentieth century, although large areas were characterized by negative trend [38–40]. Since the report of Intergovernmental Panel on Climate Change [47] raised this question that "Has the climate become more variable or extreme?" Analyzing trends in climate extreme parameters have received increasing attention from many researchers in a variety of climatological and hydrological studies. Several previous studies concerning long-term climatologically trends have focused on surface air temperature and precipitation. For example, Lettenmaier et al. [48] analyzed trends in precipitation, over the continental USA by applying the Mann–Kendall test, the results of his research showed an increase in precipitation during autumn in a quarter of the entire stations. Increasing trend in precipitation was reported by some other researchers in Australia and New Zealand [49, 50] and Argentina [51]. On the other side, decreasing trend in precipitation was the result of research which was found in the Russian Federation [52], China [53], Turkey [54], and Africa [55, 56].

Heino et al. [57] have distinguished that there is an increasing trend in minimum temperature almost everywhere and increasing trend in maximum and mean temperature in northern and central Europe. Over the Russian Federation, Canada [58] and in Australia and New Zealand an increasing trend in maximum and mean temperature was also observed [49]. These results confirm the outcomes of Smit et al.'s investigation [59] which indicate that mid-latitude regions such as the mid-western USA, southern Europe and Asia are becoming warmer and drier, whereas the lower latitudes are becoming warmer and wetter.

IPCC [38–40] reported that, during recent decades, precipitation has tended to increase in mid-latitudes, decrease in the Northern Hemisphere subtropical zones, and increase generally throughout the Southern Hemisphere. Lettenmaier et al. [48], Türke [54], Zhang et al. [60], Gonzalez Hidalgo et al. [61], Gong et al. [62], del Rio et al. [63], and Partal and Kahya [64] are some examples of the researchers who have investigated recent rainfall trends and confirm the IPCC claim about increasing and decreasing precipitation over the mentioned locations. Extreme temperature series have also received increased attention during the last decades of the twentieth century [65, 66].

The research analysis on temperature records across the world indicates there has been an increase in the mean global temperature of about 0.6 °C since the start of the twentieth century [67–69]. Mearns et al. [70] and Hansen et al. [71] concluded that small change in the mean temperature could produce substantial changes in the frequency of the extreme temperature. Increase in temperature trend of the planet has been particularly observed since 1920s [72].

According to IPCC and recorded investigations of previous researches, many regions over Europe are vulnerable to climate change impacts. The updated report of European environment agency (EEA) and intergovernmental panel on climate change (IPCC) confirms that the warming trend in Europe was above the global average mentioned since pre-industrial times. During the twentieth century, most of the Europe experienced increasing trend in average annual surface temperature with the stronger warming in winter which was observed over most regions. The 1990s were the warmest in the instrumental record.

According to IPCC [38–40], there was a 10–40 % increasing trend in precipitation in the twentieth century in northern Europe but, on the other hand, there was up to a 20 % decreasing trend in the southern part of Europe. IPCC also realized a global rising in temperature from 1.8 to 4.0 °C (3.24–7.2 F) in the twentieth century. This Projection over Europe also showed a 1.0 and 5.5 °C (1/8–9.9 F) increasing trend in temperature. This report also emphasized that there was more frequent, intense, and increasing number of hot extremes and a decreasing number of cold extremes over the past 50 years, and these trends are projected to continue.

According to updated information of Jones and Moberg [73] the warming trend between 1901 and 2005 throughout Europe was established at +0.90 °C and precipitation trends were more spatially variable. Table 2.2 shows trend analysis throughout Europe between 1977 and 2000.

Table 2.2 Trend analysis results throughout Europe between 1977 and 2000

Location	Trend	Source
Throughout Europe	Increasing trend in temperature about +0.90 °C	[73]
Central and north-eastern Europe and in mountainous regions	Increasing trend in temperature is higher. Higher trend of daily temperature due to an increase in warm extremes, rather than a decrease of cold extremes.	Böhm et al. [136]; Klein Tank and Können [137]; [73, 136, 137]
Mediterranean region	Increasing trend in temperature	Böhm et al. [136]; Klein Tank and Können [137]
In most parts of the continent	Lower trend in mean precipitation per wet day.	[136, 138, 139]
In most of Atlantic and northern Europe	Increasing in mean winter precipitation	[136]
Mediterranean area	Negative trend in annual precipitation in the east, while they are non-significant in the west.	[140]

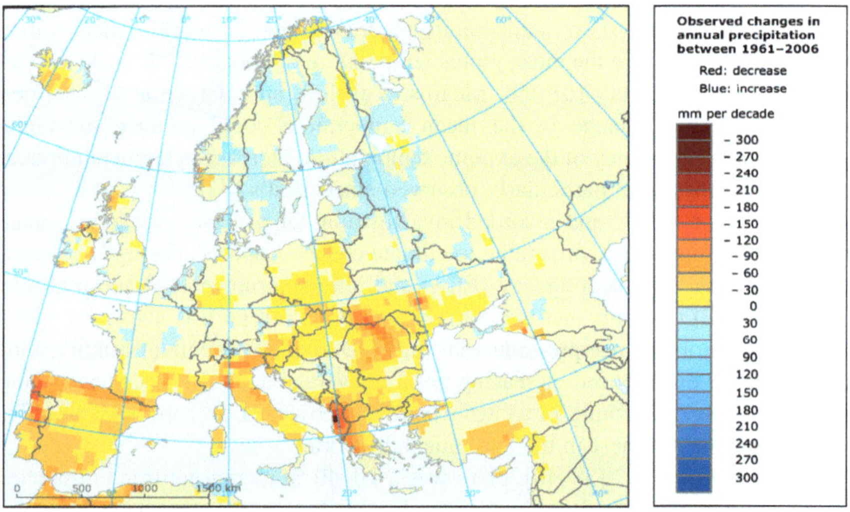

Fig. 2.3 Observed change in annual rainfall 1961–2006 (*Source* EEA's "Global and European temperature" core set indicator, based on gridded data from CRUTEM3, climatic research unit and KNMI's climate explorer)

Figures 2.3 and 2.4 show observed changes in annual precipitation and temperature between 1961 and 2006 over Europe, which was reported by EEA.

According to the robust result of IPCC [38–40], a downward trend in summer precipitation in southern Europe, accompanied by upward trend in temperatures, which enhance evaporative demand, would inevitably lead to reduced summer soil moisture [74]. Figure 2.4 shows the European mean annual temperature deviations

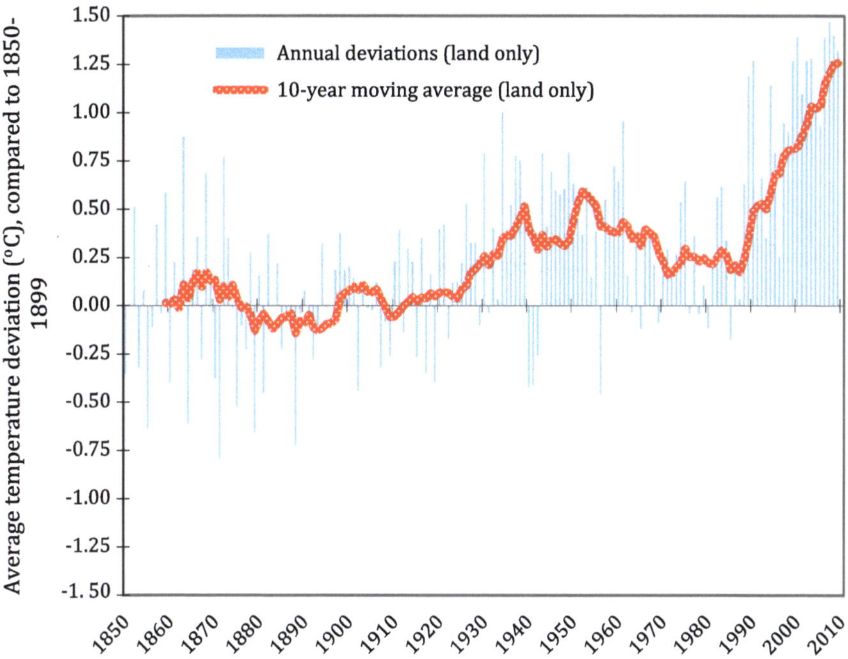

Fig. 2.4 European annual average temperature deviations, 1850–2009, relative to the 1850–1899 average (in °C). The lines refer to 10 year moving average, the bars to the annual "land only" European average; (*Source* EEA's 'global and European temperature' core set indicator, based on gridded data from CRUT)

which are in the source in relation to the base period between 1961 and 1990. To better monitor the EU objective not to exceed 2 °C above pre-industrial values, the annual deviations shown in the chart have been adjusted to be relative to the period between 1850 and 1899. Over Europe, average annual temperatures during the real pre-industrial period (1750–1799) were very similar to those during 1850–1899.

Figure 2.5 shows the 10 year moving average of the annual, winter (December, January and February) and summer (June, July and August) mean temperature deviations in Europe 1860–2009. It seems that since 1990, the annual, summer and winter variables trend in temperature over Europe was increasing. Observed temperature change over Europe 1976–2006 is shown in Fig. 2.6.

In all figures, the increasing trend in temperature over Europe is obviously clear. The increases for winter in northern Europe and in summer for southern Europe were higher.

The IPCC report also mentioned that Europe's climate was already being affected by warming in several ways; for example, the snow cover has decreased by 1.3 % per decade in the last 40 years.

According to IPCC [75], WMO [22] and McCarthy et al. [76] the strongest potential impact of climate change on human settlements is believed to be flooding [75, 22], McCarthy et al. [76]. Climatic parameters are one of the most important

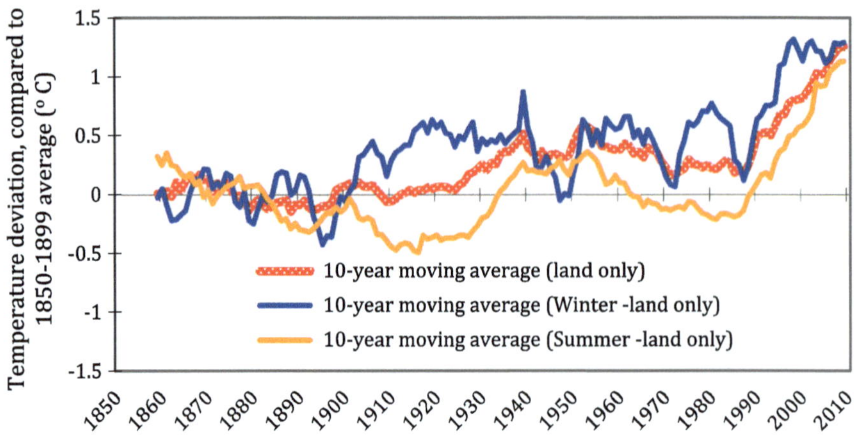

Fig. 2.5 Annual, winter (December, January, February) and summer (June, July, August) mean temperature deviations in Europe, 1860–2009 (°C). The lines refer to 10 year moving average European land; (*Source* EEA's "global and European temperature" core set indicator, based on gridded data from CRUTEM3, climatic research unit and KNMI's climate explorer)

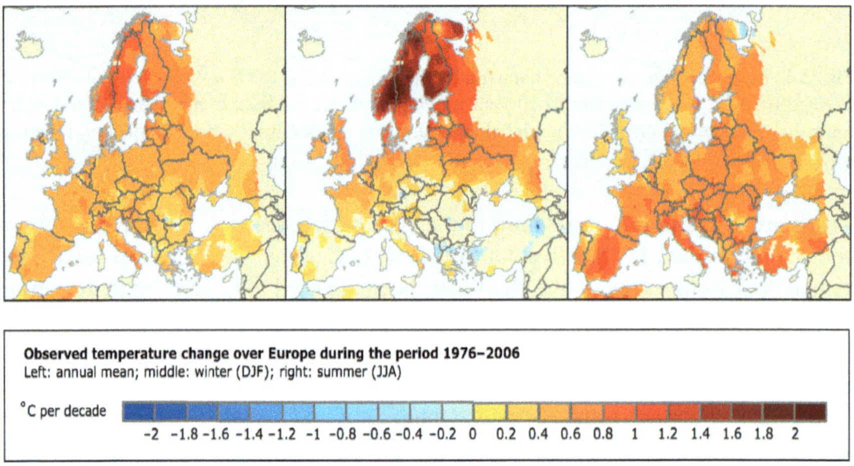

Fig. 2.6 Observed temperature change over Europe 1976–2006; (*Source* EEA's "Global and European temperature" core set indicator, based on gridded data from CRUTEM3, Climatic Research Unit and KNMI's climate explorer)

triggering factors for floods event. The trend analysis of hydrological and meteorological series is important, even more relevant when considering the regional effects of global climate change [22, 77]. As the atmosphere's water holding capacity increases with temperature, the intensified potential for extreme precipitation events augments the risk of inundations caused by sustained rainfall over most land areas [39, 78], especially in areas where flooding is typically triggered by intense summer rain [33]. Alternatively, decreases in snow and river-ice cover reduce the frequency and severity of snowmelt-related and ice-jam floods [79–83]. To date, relatively few researches have made a quantitative assessment of the potential impacts of climate change on extreme river flows in Europe. These studies were applied in some regions by different scientists; UK: Kay et al. [84]; Scandinavia: Graham et al. [85]; the Benelux countries: Booij [86]; and Germany: Shabalova et al. [87]. While some of the investigators realized an increasing trend in flood frequency and intensity, the other investigators found a decreasing trend.

Application of different climate scenarios as well as hydrological models made it difficult to compare the results and to make a picture at a European scale. Lehner et al. [88] analyzed the changes in flood frequencies due to global climate change over Europe, and found northern to north-eastern Europe to be mostly affected by global climate change, and distinguished an increasing flood risk in those areas. However, their investigation was based on applying the climate change signal of two different general circulation models (GCMs), not based on an observation-based dataset. They also did not take into account a potential increase in climate variability [88].

Additionally, from natural features point of view in the climate systems, the global climate change is also related to anthropogenic influences [89, 90]. The effect of human beings on hydrological time series therefore should be of great attention around the world. The example of increasing temperature and decreasing precipitation since the mid1960s as a result of deforestation was presented by Kothyari and Singh [91]. Similarly, Meher-Homji [92] showed decreasing precipitation trends because of increasing deforestation. Sharma et al. [90] also showed some evidence of increasing temperature and decreasing precipitation and discharge particularly during low flow season, as a result of land use change, and anthropogenic effects.

2.7 Climate Change Scenario for Barcelonnette Area

For Barcelonnette area, there are only a few studies about estimation of climate change scenarios with focusing on landslide. In one example, influence of climate change scenario on slope hydrology and landslide frequency is investigated by Malet et al. [93].

They applied climate change scenarios of GCMs and analyzed a future scenario for the period of 2069–2099 based on observed data between 1969 and 1999. The results of impacts of climate change based on the A2 scenario are shown in the Fig. 2.7. According to Malet et al. [93], the main results from this investigation for Southeast France ere: (a) higher temperatures in summer; (b) more rainy winters; (c) drier summers; and (d) a decrease in soil water content.

In another example, Buma and Dehn [94] investigated also the prediction of climate change impact on slope stability using downscaled climate data and slope hydrology/stability model. They applied three scenarios for the period of

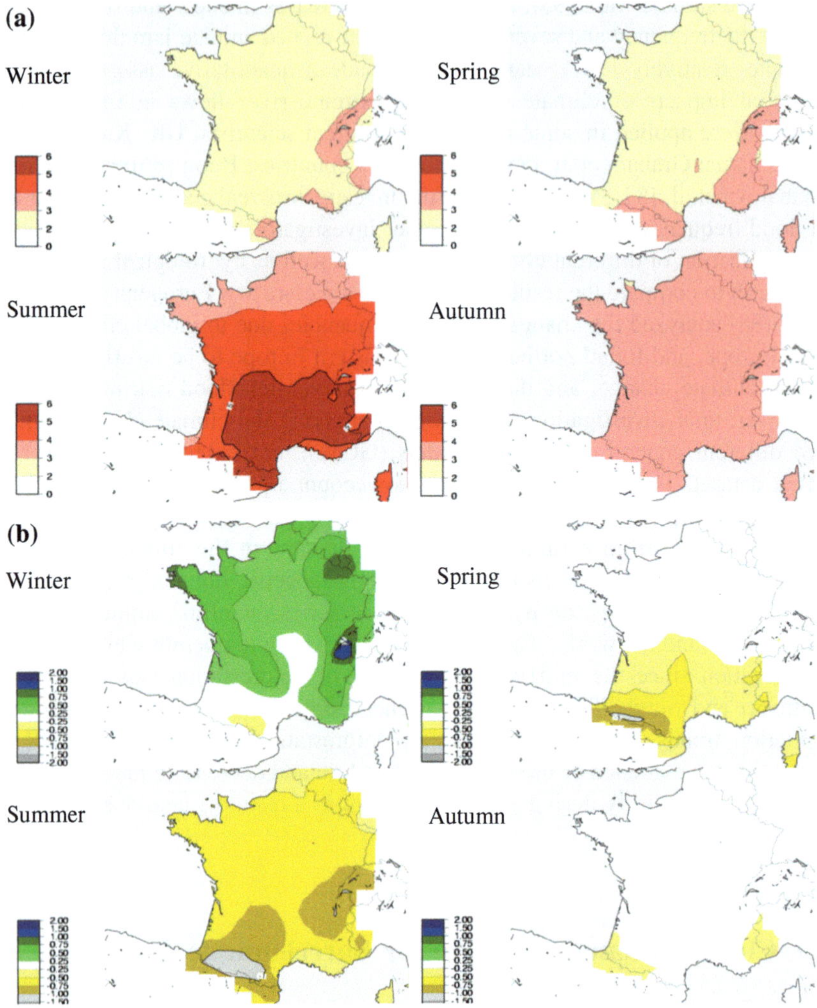

Fig. 2.7 Impacts of climate change based on the A2 scenario of GIECC. **a** Changes in temperature. **b** Changes in precipitation [93]

1971–2000; 2021–2050 and 2069–2099 according to the observed data between 1928 and 1970. In this research, they found a decreasing in local winter precipitation in Barcelonnette area. Buma and Dehn [95] in another study investigated the effects of climate change on landslide in South East France. They found that precipitation increased in winter and spring, and decreased in summer and autumn, using the reference period of analog-downscaling with HCGS between 1950 and 1979.

2.8 Land Use Change and Anthropogenic Factors

Apart from climatic factors that have affected floods (as mentioned in Sect. 2.6), according to Kundzewicz and Schellnhuber [96] another significant reason, for increasing flood hazards can be related to nonclimatic anthropogenic factors: Changes in socioeconomic systems include land-use changes such as deforestation, urbanization, elimination of floodplains area, as well as river regularizations [96], which lead to reducing of water storage, reducing of infiltration capacity, and increasing of runoff coefficient [97]. WMO [22] also reported that the first category of nonclimatic anthropogenic factors effects resulting from human actions which have affect on the ecosystem changes are deforestation and urbanization. These lead to changes in the ecosystem that magnify the consequences of heavy precipitation, converting this precipitation into floods of a greater severity than otherwise would have resulted [22].

Xiaoming believes that land use change is a major reflection of the land ecosystem change, which will then lead to the hydrology regime changes. The hydrology regime changes can then feedback on environmental and land use change, which generates a complex and international system.

By floodplain development, increasing accumulation of population and wealth in flood-prone areas, humans have been driven to occupy unsafe areas (e.g., informal settlements on floodplains), thereby increasing the loss potential.

Where urbanization occurs on the floodplain, the ability of the floodplain to attenuate a flood peak by promoting storage, infiltration, and alternative flow pathways is reduced. This effect is greatly enhanced where flood defenses are erected; such structural measures may also cause residents to lose their sense of natural river dynamics and reduce the perceived risk for further development [98].

The history analysis of urban growth indicates that urban areas are the most dynamic places on the earth's surface. Despite their regional economic importance, urban growth has also a considerable impact on the surrounding ecosystem changes [99]. In the last few decades, an increase in urban areas has occurred in the world, and demographic growth is one of the major factors responsible for these increasing changes. By 1900 only 14 % of the world's population was residing in urban areas, but it had increased to 47 % by 2000 [100]. Urban growth is a common phenomenon in almost all countries over the world. Currently, increases in urban areas are the major environmental concerns that have to be analyzed and monitored for any future planning.

Urbanization and its increasing trend may have particularly severe influences on small catchments, where a high percentage of the catchment area may undergo a change in land use within a short time period. In a larger catchment, the effects of land use change would to a greater extent be damped by the remainder of the catchment area [101–103]. Since the end of the 1960s, the scientific literature reported the results of studies on the possible effects on the fluvial regime of intense deforestation and urbanization which had occurred in some drainage basins of the United States [104]. Some previous studies also have analyzed the effects of intensive land use change, such as the construction of roads close to the rivers in mountainous areas (as a type of land use change and anthropogenic activity), can induce on the river runoff which then increase [105].

These studies highlighted that the hydrological effects of human activity are strictly dependent on the extension of the area affected by the man made intervention. For instance, similar analyses were performed in a tropical watershed by Costa et al. [106]; he realized that a deforestation of about 30 % in the basin induced a 24 % increase in annual mean discharge. Therefore, it might be concluded that land use change, in particular deforestation, has been credited with causing important increases in the frequency and severity of flood events. In another simulation, the research performed by Brath and Montanari [107], the higher sensitivity to land use change of the low flows has been also investigated. Recently, Naef et al. [108] reached similar results and conclusions by addressing a river basin located in Germany. They found that the flood runoff reduction by land use change can be remarkable in the presence of rapid run-off production only. In the last two decades, a lot of research has been carried out to estimate the effects on the hydrological cycle induced by vegetation cover changes, with particular emphasis on deforestation consequences [109], urbanization of bottom valley areas [110], construction of roads in forests [111, 112], and conversion of wooded areas to pasture [113]. In recent decades, the scientific investigation shows that a lot of inundation has occurred in Europe causing loss of human lives and financial damages which have been aggravated, in several cases, by the intense urbanization of flood-prone areas [114]. Based on the European parliament's report, land-use changes in Europe have lead to a reduction in the storage volume and an increase in the run-off coefficient. Sullivan et al. [115] studied a basin in Cornwall which had a significant increasing trend in magnitude and frequency of flood flows, but only a weak decreasing trend in rainfall. Although they suggested that the increasing in flood flows could be attributed to land-use change. Other researchers have also suggested that this interaction of different forcing factors could be exploited, for example by offsetting urbanization with forestation, or climate change by land-use change [108, 116, 117, 118]. The effect of land-use change was studied also by Bultot et al. [119], who analyzed the influence of land use on the water balance of the near surface soil layer by applying a conceptual rainfall–runoff model to a Belgian river basin. They concluded that the presence of vegetation can induce effects on the river flows that are more evident in arid climates where the vegetation cover causes a reduction of the river discharge that is more marked for the lower river flows. Increases in urban areas tend to increase the responsiveness of an area to a rainfall event, usually leading to flash flooding and increased maximum rates of stream flow. Infrastructure planning and implementation, as part of a regional development plan of an urban area, contributes greatly to mitigate damages of flood events. It should be considered that while infrastructure planning and implementations reduce flood damage from small and medium floods; they can induce catastrophic floods when they start to fail. For example, if dykes or levees fail, they can cause a false sense of security on public and residential parts in flood-plain areas. As noted by Eiker et al. [120] for flood-mitigation projects, the question is not if the capacity will be exceeded, but what are the impacts when the capacity is exceeded. Thus, land-management planners and the public must fully understand the consequences when the dykes or dam fails.

2.9 Historical Changes in Land Use in Barcelonnette Area

One of the reasons for deforestation during the seventeenth century in France was that the forests were considered as an economic resource spatially for the consideration of warships. The dramatic consequences of deforestation led to increasing of land degradation and mass movement occurrence [121]. Many authors have addressed in their investigations to deforestation in the floodplains of the rivers in south-eastern France in the late eighteenth and nineteenth centuries [122]. In contrast, with the vegetation explosion of the early twentieth century, changes were observed over shorter time periods and in smaller areas, in the riparian forest active channel contacts. Table 2.3 summarizes previous investigations on land use, channel morphology and population change during eighteenth, nineteenth, and twentieth centuries in whole catchment:

Bravard [123] by investigation on previous research concluded that anthropogenic factors played a major part in initiating active channel restriction at the turn of the century, with climate change as a secondary factor. Slopes were stabilized through extensive reforesting, torrent control measures, and progressive abandonment of agro-pastoral activities, thereby reducing peak flow and bed-load supply. Bravard [123] also confirmed the importance of natural and anthropogenic factors in riparian vegetation in his research. The evolution of the Ubaye Riverbed at the turn of the twentieth century provides an instructive example of short-term watershed influence on a river segment. As shown by Schumm [124], channel geometry is equally adjusted to external factors and is modified at the same time as riparian vegetation. Channel deepening and narrowing on the major part of the Ubaye course was noted, as well as channel pattern modification. By increasing hydraulic roughness and favoring bar stabilization, forest expansion exerts an internal control on the alluvial mosaic and contributes to a reduction in bed width. The genesis of a floodplain forest is, in fact, accompanied by the development of tree units within the active channel, and therefore rivers draining forest corridors are generally not as wide as those located in prairie sectors [125]. Internal factors,

Table 2.3 Summarizing historical land use change

Years	Changes	Source
1830	In 1830, the Ubaye river occupied most of the valley bottom, whereas agricultural land use extended up to the edge of the active channel.	[141]
Since 1860	The mobility of channel-forest contact the cyclic evolution of the forest margin-active channel contact during the recent period appears to be controlled by fluctuations in hydrology and bed-load. The population decreased by 50–60 % in the Barcelonnette basin and by more than sixfold in the upper valley. Since 1860, when the population peaked in this region, the population has decreased by 75–80 %.	[141]
As early as 1892–1905	During this time 3,500 ha were reforested with Austrian black pine while 8,589 ha had already been reforested (Figs. 2.8, 2.9)	[142]
1920–1925	In 1920 Riparian forestation started. The riparian forest development in the active channel between 1920 and 1925 was the result of reduced rejuvenation processes like peak flow and bed load supply, mainly due to a slope afforestation policy and torrent correction works undertaken in the late nineteenth century by the RTM agency (law of 28 July, 1860).	[141]
1920–1930	The active channel area was colonized by Pinus Sylvestris. Dendrochonological analysis shows that forest recolonization of the active channel took place around 1920–1925 based on age of Pinus Sylvestris. Recolonization was accompanied by fluvial metamorphosis, changing from a braided pattern to a sinuous single-bed pattern, as well as by bed incision. The topographical cross-section of this sector demonstrates that the actual bed is 2 m lower than it was 75 years ago.	[141]
1941–1957	Close chronological ties with active channel expansion (+12.3 % from 1948 to 1956).	[141]
Between the first third of the nineteenth century and 1948	The area of active channel decreased from 246 to 252 ha to 174 ha.	[141]
1948–1973, notably between 1948 and 1956	Vegetative colonization of the active channel primarily affected the segment upstream from Barcelonnette where the forest decreased by 44 %. Such decreases were also occasionally observed downstream, but not in any case throughout the entire period. Reduction of forest area in favour of the active channel area. These changes, lasting only two decades, were in opposition with the trend observed at the beginning of the century.	[141]

(continued)

Table 2.3 (continued)

Years	Changes	Source
After 1950	Spatial fluctuation of forest-channel borders occurred.	[141]
1957	Flood 1957 impacts had been largely attenuated below the braided section upstream from Barcelonnette. The exceptional dimensions of the floodplain in this sector greatly contributed to attenuating the flood, favoring bed load deposition. Thus, the millennial flood caused comparatively little damage to the riparian vegetation in the study area, excepting the upstream sector which is characterized by specific changes. Conversely, dykes downstream were destabilized by a 0.5–1 m bed incision. The 1957 flood also reduced channel slope and hydraulic power upstream from Barcelonnette, which explains the adjustment of vegetation and bed geometry over the ensuing decades. These adjustments are ongoing at this site, whereas they have not been observed since 1982 downstream from Barcelonnette.	[143]
1948–1990	In fact, sector analysis from 1948 to 1990 shows a complex geographical evolution. The forested areas were almost stable as tree units went from 209 ha to 196 ha show a slight reduction of 6 % (Fig. 2.10). The active channel was also restricted during this period mostly due to human installations developed in the corridor downstream from Barcelonnette.	[141]
1970–1990	In contrast the 1970–1990 periods is characterized by reduced peak flow and vegetative recolonization (+5 %) of the active channel (Fig. 2.11).	[141]
The 1973–1990	The sector located upstream from Barcelonnette showed major continuous vegetation colonization of the active channel. Whereas downstream, these changes were discontinuous and were only apparent between 1973 and 1982. The forest progression and additional restriction of the active channel (Fig. 2.12). The forest gained only 9 ha, whereas the active channel lost 40 ha, as a result of human modification (water treatment stations, camping grounds, and public dump sites).	[141]
1982 and 1990	A great part of the active channel was colonized by pioneer units. These units were observed in the 1982 aerial photographs, (+19 ha).	[141]
Early twentieth century	Changes were observed in the riparian forest active channel contacts over shorter time periods and in smaller areas. Forest covered only 10.3 % of the watershed in the early nineteenth century, but from twentieth century, occupies 33 % of the watershed area. This recolonization has affected all the local districts, as 70 % of them have more than doubled in forested area (Figs. 2.13, 2.14).	[141]

(continued)

Table 2.3 (continued)

Years	Changes	Source
At the end of the nineteenth century, during the twentieth century	The reach located upstream from Barcelonnette was characterized by a dynamic equilibrium corresponding to high peak flow and abundant bed-load. This segment is typical of alpine fluvial landscapes which registered major modifications between the late fourteenth century and the nineteenth century, evolving toward active braiding. While for many years mountain populations were considered to be responsible for this evolution, recent views have focused on the influence of the little ice age climatic degradation, affecting slopes whose vegetation cover was altered by overgrazing. This led to an intensification of flooding and to an increased bed-load supply which, in turn, produced channel aggradations and widening. Thus, the Ubaye occupied the entire valley flat when human society exploited slopes to the fullest, by combining dominant pastoral activities with subsistence crops.	[144, 20, 145]

Fig. 2.8 An evolving land use: intense deforestation in the eighteenth century (agro-pastoral communities); intense gullying (*source* RTM)

Fig. 2.9 An evolving land use: intense deforestation in the nineteenth century (agro-pastoral communities); intense gullying (*source* RTM)

which are biological or physical, are in fact interrelated and adjusted with the external factor changes. This afforestation is different from situations observed downstream of flood control dams [126]. Its evolution is more progressive and more long term, as riparian vegetation eventually adjusts to natural hydrological variations. This phenomenon also differs from other cases of piedmont river evolution where vegetation metamorphosis has been found to be due to floodplain abandonment by farmers [127].

Fig. 2.10 An evolving land use: intense deforstation in the nineteenth century (agro-pastoral communities); intense gullying (*source* RTM)

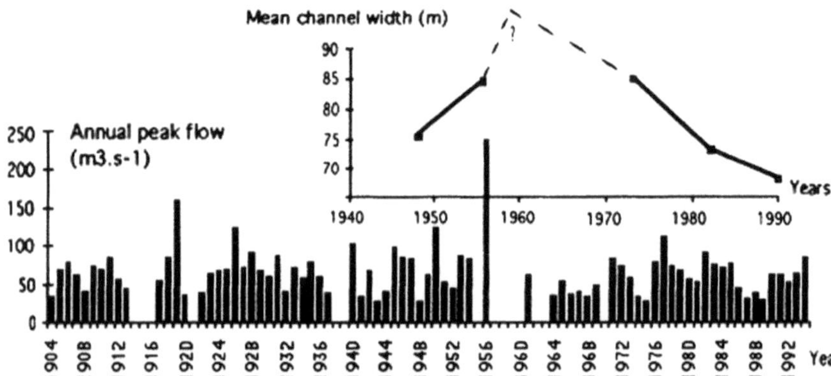

Fig. 2.11 Evaluation of active channel average width and annual peak flow from 1945 to 1990, on the middle Ubaye (After Hydro bank data and aerial photographs) (*source* RTM)

Fig. 2.12 An evolving land use: intense deforestation for torrential control during the twentieth century: 1987 (*source* RTM)

Fig. 2.13 An evolving land use: during the twentieth century, intense reforestation for torrential control (*source* RTM)

Fig. 2.14 An evolving land use: during the twentieth century (*source* [93])

2.10 Physical Characteristic of Flood Hazard

In the frame of hazard assessment, flood events are analyzed by means of recurrence intervals and spatial–temporal flood characteristics [128]. Flood hazard estimation is based on the factors such as the triggering factors causing the hazard, their spatial extent, duration and time of onset, including their frequency and magnitude of occurrence, and secondary events influencing the event if any [129]. The flood characteristic creates a clear understanding of the flood plain, and gives the understanding to the flood hazard behavior. The flood characteristic is not only described by flood inundation, but also by inundation depth, flow velocity, inundation duration; volume of water, surface area, stream power, and so on. Access to all this information is possible by flood simulation using hydrodynamic modeling.

Alkema [130] emphasized that the hazard estimation is also based on the output of the model simulation in the form of parameter maps such as flood velocity, depth of water, and flood impulse. The pre-requisite for flood hazard is estimated by its frequency analysis or return period calculation [131]. Hydrographs for different return periods are the basis for understanding the hydrologic response of the basin [132].

The following characteristics are important in terms of the physical hazard posed by floods event:

a. The depth of water and its spatial variability;
b. The extent of inundation, and in particular the area that is not normally covered with water;

c. The water velocity and its spatial variability;
d. Duration of flooding;
e. Capacity for erosion and sedimentation (In this research this parameter did not conclude due to lack of data)

The importance of water velocity should not be underestimated, as high velocity water can be extremely dangerous and destructive. In the case of a flood flowing into a reservoir, the flood volume and possibly hydrograph shape should be added to the list of important characteristics. In most cases, however, the flow rate is important because it is used, in conjunction with the topography and condition of the channel/floodplain in determining the water depth, velocity, and area of inundation. Characteristics such as the number of rivers and streams involved in a flood event, total size of the affected area, duration of flooding, and the suddenness of onset are related to the cause of flooding. Usually, these space–time factors are determined primarily by the space–time characteristics of the causative rainstorm and secondarily by watershed characteristics such as area and slope. Because of the seasonality of flood-producing storms or snowmelt, the probability of floods occurring in a given watershed can differ markedly from season to season. On a given river, small floods (with smaller discharges, lower stages, and limited aerial extent) occur more frequently than large floods. Flood-frequency diagrams are used to illustrate the frequency with which floods of different magnitudes occur. The slope of the flood-frequency relation is a measure of the variability of flooding [22].

The development of the flood hazard map, and its subsequent reviews, will be carried out to include as much information to assist future researches. The flood hazard maps in this thesis cover the geographical areas which could be flooded according to the following scenarios:

a. Floods with a low probability, or extreme event scenarios;
b. Floods with a medium and high probability (return period $\leq$100 years);
c. Floods with a high probability (with the assumption of dyke failure).

For each scenario referred the following elements are shown:

a. The flood extent;
b. Water depths or water level;
c. Where appropriate, the flow velocity or the relevant water flow.

Simulation and modeling for flood estimation is a rapidly developing field in hydrology. The flood simulation and model results provide the authorities with relevant information on how the flood is going to behave at the location where people live and how the flood will affect them [133].

2.11 Presentation of Hazard Assessment

In order to have available and more effective tools for providing the people with information, as well as a valuable basis for priority setting and further environmental, technical, financial, and political decisions regarding flood hazard/risk management, it is necessary to establish flood hazard maps. These maps are showing the potential adverse consequences associated with different flood scenarios, including information on potential sources of environmental damages as a consequence of floods [37, 134].

Assessment of flood hazard is extremely important in hydraulic design, river engineering, and land management, e.g., the construction of buildings and residence is often restricted in high flood hazard areas and flood plain. Critical facilities such as hospitals or nuclear power plants should be constructed and located in low-flood-hazard areas or in areas where the flood hazard is essentially zero [135]. For locations where dam failure may occur in case of massive flooding, dam spillways must be fit designed to pass extremely large floods without dam failure occurrence [22].

Maps are the standard format for presenting flood hazards. Areas subject to flooding are indicated on topographic base maps through shading, coloring, or drawing lines around the indicated area. The flood-hazard areas may be divided according to severity (deep or shallow), type (quiet water or high velocity), or frequency of flooding. Different symbols (different types of shading, colors or lines) should be used to clearly indicate the different types of flood-hazard area, and there should be written explanations, either on the map or in an accompanying report, as to the exact meaning of the symbols. The maps will be easier to read if extraneous information is omitted from the base maps. Maps should always have a graphic scale. Numeric scales (e.g., 1:1 000) lose their validity when the map is reduced or enlarged. Ancillary information may accompany the basic maps: flood-frequency diagrams; longitudinal profiles or channel cross-sections showing water level as a function of flood frequency; information on velocity; suddenness of onset; duration of flooding; the expected causes; and season of flooding.

The actual maps can be prepared manually using standard cartographic techniques or with a GIS. The format and scale of a hazard map will depend on the purpose for which it is used, and it may be desirable to have more than one type of map. High-resolution flood maps are necessary to show the exact location of the flood hazard. Such maps may be used by individuals and authorities to direct new construction into relatively safe areas. For purposes of disaster preparedness, planning and relief efforts, it is best to have maps which depict all types of hazards (natural and human induced).

Disaster-planning maps should also show population and employment centers, emergency services and emergency shelters, utilities, locations of hazardous materials, and reliable transportation routes. It is useful to show which bridges and roads are likely to be made impassable by flooding of various magnitudes, and which are likely to be passable under all foreseeable conditions. Even if a disaster-response plan has not been formulated, these maps can be used in the event of a disaster to direct relief to critical areas by the most reliable routes. Photographs are one of the most effective ways of communicating the consequences of a hazard. If photographs that are appropriate to the local nature of the hazard accompany hazard maps, then more people are likely to pay attention to them. Communication of the infrequent and probabilistic nature of the hazard is important, though difficult. This is particularly important in areas protected by levees. Hazard maps should be made widely available in paper format to local communities and authorities. They should be distributed to:

a. Those who may be involved in disaster-relief efforts;
b. The public; and
c. Those who may be in a position to implement mitigation measures.

For planning and evacuation procedures, the demand for flood information and digital maps has been increased. Ideally, the key organizations involved in disaster-relief efforts will have the maps displayed permanently on a wall, and will have studied the maps and instituted disaster planning. Ideally, the public, community leaders, and government bodies will also study the maps and appreciate that prevention is worthwhile, and implement appropriate mitigation measures. Also, near full-scale disaster exercises may be conducted periodically to maintain the readiness of disaster relief and management organizations, and to keep the public aware of the potential hazard [22].

2.12 Chapter Summary

This chapter gave a comparative overview of the major challenges faced when dealing with hydrodynamic simulation and flood hazard. In the first part, definition, source, and implication of digital elevation model were explained. DEM is usually produced form sampled or observed data. Spurious sinks or local depressions in DEMs are frequently encountered and are a significant source of problems in hydrological applications. In this case, application of interpolation techniques seems to be necessary to correct DEMs. Therefore, in this part, the attempt was introducing different interpolation techniques drainage network estimation coupled with advantages and disadvantages of these methods and their application in different research fields. In the next part, flood hazard, concepts, definition, types, and causes were defined. Then, the focus was on providing a background to climate change and land use change as the most important triggering factors which affects on flood events. This background was overviewed and estimated from global to local (Barcelonnette area) based on different scientific researches. According to previous researches, there have been adverse flood hazard changes due to climate change. Many investigators believe that climate change is expected to have substantial impacts on hydrology on global, regional and local scale. According to IPCC, increasing trend in temperature was distinguished about +0.90 °C in Europe.

In some other researches negative trend in annual precipitation was found in the east of Europe. IPCC [38–40] also reported that there was a 10–40 % increasing trend in precipitation in the t century in northern Europe but, on the other hand, there was up to a 20 % decreasing trend in the southern part of Europe. Apart from climatic factors, another significant reason, for increasing flood hazards can be related to non-climatic anthropogenic factors: Changes in socioeconomic systems include land-use changes such as deforestation, urbanization, elimination of floodplains area, as well as river regularizations which leads to reducing of infiltration capacity and increasing of runoff coefficient. In recent decades, the scientific investigation shows that a lot of inundation has occurred in Europe causing loss of human lives and financial damages which have been aggravated, in several cases, by the intense urbanization of flood-prone areas.

For Barcelonnette area, the effect of climate change on landslide and slop stability using downscaling methods was investigated by some researchers. The general results were higher temperatures in summer, more rainy winters and spring, drier summers with decreasing precipitation, and a decrease in soil water content. In terms of land-use change, many authors have addressed in their investigations to deforestation in the floodplains of the rivers in south-eastern France in the late eighteenth and nineteenth centuries ([122]. In contrast with the vegetation explosion of the early twentieth century, changes were observed over shorter time periods and in smaller areas, in the riparian forest active channel contacts. Bravard [123] realized that anthropogenic factors played a major part in initiating active channel restriction at the turn of the century, with climate change as a secondary factor in Barcelonnette area.

In the next chapter, the methodology applied in this research will be explained in details.

References

1. Vaze J, Teng J, Spencer G (2010) Impact of DEM accuracy and resolution on topographic indices. Environ Model Softw 25:1086–1098
2. Hutchinson MF, Gallant JC (2000) Digital elevation models and representation of terrain shape. In: Wilson JP, Gallant JC (eds) Terrain analysis. Wiley, New York, pp 29–50
3. Hutchinson M, Stein J, Stein J (2008) User's guide, GEODATA 9 second "DEM and D8", digital elevation model version 3 and flow direction grid, fenner school of environment and society. Australian National University (ANU) and Hamish Anderon and Phil Tickle, geosciences Australia
4. Teng J, Vaze J, Tuteja NK, Gallant J (2008) A GIS based tool for spatial and distributed hydrological modelling: CLASS spatial analyst. Trans GIS 12(2):209–225
5. Burrough PA, McDonnell RA (1998) Principals of geographical information systems. Oxford University Press, New York, p 333
6. Wood J (1996) The geomorphological characterization of digital elevation models. Ph.D. Dissertation, Department of Geography, University of Leicester, Leicester, UK
7. Olivera F, Furnans D, Maidment D, Djokic Z (2002) Drainage system, Arc hydro: GIS for water resources. In: Maidment DR (ed) ESRI Press, Redlands, p 203

8. Wechsler S (1999) Results of the DEM user survey. Available online at: http://web.syr.edu/~sperlit/survey. Verified 28 Nov 2011
9. Harding DJ, Bufton JL, Frawley J (1994) Satellite laser altimetry of terrestrial topography: vertical accuracy as a function of surface slope, roughness and cloud cover. IEEE Trans Geosci Remote Sens 32:329–339
10. Burrough PA (1986) Principles of geographical information systems for land resources assessment, vol 12. Clarendon Press, New York
11. Hutchinson MF, Dowling TI (1991) A continental hydrological assessment of a new grid-based digital elevation model of Australia. Hydrol Process 5:45–58
12. ANUDEM User's Manual (2011) Fenner school of environment and society. ANU College of Medicine, Biology and Environment. http://fennerschool.anu.edu.au/research/publications/software-datasets/anudem. Verified 28 Nov 2011
13. Jenson SK, Domingue JO (1988) Extracting topographic structure from digital elevation data for geographic information system analysis. Photogram Eng Remote Sens 54:1593–1600
14. Taylor J (1997) An introduction to error analysis: the study of uncertainties in physical measurements. University Science Books, Sausalito, p 327
15. Podobnikar T (2006) DEM from various data sources and geomorphic details enhancement, 5t ICA mountain cartography workshop. Bohinj, 31 Mar 2006
16. Fürst J (2004) Tutorial: interpolation of hydrological variables. University of Natural Resources and Life Sciences, Vienna
17. Smith MJ, Goodchild MF, Longley PA (2009) Geospatial analysis, a comprehensive guide to principles, techniques and software tools, 3rd edn., Issue version: 3.1 2—Published by Matador (an imprint of Troubador Publishing Ltd) on behalf of The Winchelsea Press
18. Akkala A, Devabhaktuni V, Kumar A (2010) Interpolation techniques and associated software for environmental data. Environ Prog Sustain Energ 29:134–141
19. ArcGIS 9.3-online help (2011) Available online at: http://webhelp.esri.com. Verified 28 Nov 2011
20. Wilson JP, Gallant JC (2000) Differences in topographic characteristics computed from 100- and 1000-m resolution digital elevation model data. Hydrol Process 14:987–1002
21. Wilson JP, Gallant JC (2000) Digital terrain analysis. In: Wilson JP, Gallant JC (eds) Terrain analysis: principles and application. Wiley, New York, pp 1–27
22. IAEA/WMO (2006) Global Network of Isotopes in Precipitation. The GNIP Database. (http://isohis.iaea.org), accessed August 2010
23. Wise SM (1998) The effect of GIS interpolation errors on the use of digital elevation models in geomorphology. In: Lowe SN, Richards KS, Chandler JH (eds) Landform monitoring, modelling and analysis. Wiley, New York, pp 139–164
24. Kundzewicz Z (2003) Extreme precipitation and floods in the changing world. IAHS-AISH Publ 281:32–39
25. Directive 2007/60/EC of the European parliament and of the council of 23 Oct 2007 on the assessment and management of flood risks text with EEA relevance. Official Journal of the European Union, L 288, pp 0027–0034
26. UN-ISDR (2004) International Strategy for Disaster Reduction, Available online at: http://www.unisdr.org (Verified 28 Nov 2011)
27. Aysan YF (1993) Vulnerability assessment. In: Merriman PA, Browitt CWA (eds) Natural disasters: protecting vulnerable communities. Thomas Telford, London, pp 1–14
28. Davis L (1992) Natural disasters: from the black plague to the eruption of Mt. Pinatubo, Facts on File, 321:1976
29. Wasseff AM (1993) relative impact on human life of various types of natural disaster: an interpretation of data for the period 1947–1991. In natural disasters: protecting vulnerable communities. Thomas Telford, London, pp 15–24
30. Wijkman A, Timberlake L (1984) Natural disasters: acts of god or acts of man?. International Institute for Environment and Development press, Washington, p 126

31. Yen C-L, Yen BC (1996) A study on effectiveness of flood mitigation measures. In: Maxwell WHC, Preul HC, Stout GE (eds) Proceedings, Rivertech'96, 1st international conference on new/emerging concepts for rivers, International Water Resources Association, Albuquerque, NM, pp 555–562
32. Munich Reinsurance (1997) Topics: annual review of natural catastrophes 1996, Munich Re 2005. p 16
33. Kundzewicz ZW, Jun X (2004) Towards an improved flood preparedness system in China Hydrol Sci J 49(6):941–944
34. Schultz B (2006) Floods management under rapid urbanization and industrialization in flood-prone areas: a need for serious consideration. Irrig Drainage 55:3–8
35. World Meteorological Organization (WMO) (1981a) Flash flood forecasting, operational hydrology report no. 18, (WMO-No. 577), Geneva, Switzerland, p 47
36. World Meteorological Organization (WMO) (1994) Guide to hydrological practices, (WMO-No. 168), vol 2, 5th edn. Geneva, Switzerland, p 765
37. COM (2004) 472 final, Brussels, 12 July 2004, Commission of the European communities, Communication from the commission to the council, The European parliament, The European economic and social committee and the committee of the regions, Flood risk management, Flood prevention, protection and mitigation
38. Intergovernmental Panel on Climate Change (IPCC) (2007a) Climate change 2007: the physical science basis. Contribution of working group I to the fourth assessment, Report of the intergovernmental panel on climate change. Cambridge University Press, Cambridge and New York, p 996
39. Intergovernmental Panel on Climate Change (IPCC) (2007b) Summary for policymakers. In: Parry ML, Canziani OF, Palutikof JP, van der Linden PJ, Hanson CE (eds) Climate change 2007: impacts, adaptation and vulnerability, contribution of working group ii to the fourth assessment report of the intergovernmental panel on climate change. Cambridge University Press, Cambridge, UK, pp 7–22
40. Intergovernmental Panel on Climate Change (IPCC) (2007c) Climate change 2007: synthesis report, contribution of working group to the fourth assessment report of the intergovernmental panel on climate change (IPCC). Cambridge University Press, Cambridge
41. United States Search and Rescue Task Force (2011) Floods and flash floods, available online at: http://www.ussartf.org/flooding.htm. Verified 28 Nov 2011
42. Kundzewicz ZW, Jun X (2004) Towards an improved flood preparedness system in China. Hydrol Sci J 49(6):941–944
43. Andréasson J, Bergström S, Carlsson B, Graham L, Lindström G (2004) Hydrological change—climate change impact simulations for Sweden. Ambio 33:228–234
44. Lotsari E, Veijalainen N, Alho P, Käyhkö J (2010) Impact of climate change on future discharges and flow characteristics of the Tana river, Sub-Arctic Northern Fennoscandia. Geografiska Annaler: Ser A, Phys Geogr 92:263–284
45. Kjellström E, Bärring L, Gollvik S, Hansson U, Jones C, Samuelsson P, Rummukainen M, Ullerstig A, Willén U, Wyser K (2005) A 140 year simulation of European climate with the new version of the Ross by centre regional atmospheric climate model (RCA3). Reports meteorology and climatology, 108, SMHI, SE-60176 Norrköping, Sweden, p 54
46. Beldring S, Engen-Skaugen T, Førland EJ, Roald LA (2008) Climate change impacts on hydrological processes in Norway based on two methods for transferring regional climate model results to meteorological station sites. Tellus A 60:439–450
47. Intergovernmental Panel on Climate Change (IPCC) (1995) Climate changes 1995: contribution of working group I to the second report of the intergovernmental panel on climate change. pp 141–193
48. Lettenmaier DP, Wood EF, Wallis JR (1994) Hydro-climatological trends in the continental United States (1948–1988). J Clim 7:586–607

49. Plummer N, Salinger MJ, Nicholls N, Suppiah R, Hennessy KJ, Leighton RM, Trewin B, Page CM, Lough JM (1999) Changes in climate extremes over the Australian region and New Zealand during the twentieth century. Clim Change 42:183–202
50. Suppiah R, Hennessy K (1998) Trends in total rainfall, heavy rain events and number of dry days in Australia, 1910–1990. Int J Climatol 10:1141–1164
51. Viglizzo EF, Roberto ZE, Filippin MC, Pordomingo AJ (1995) Climate variability and agroecological change in the central Pampas of Argentina. Agric Ecosyst Environ 55:7–16
52. Gruza G, Rankova E, Razuvaev V, Bulygina O (1999) Indicators of climate change for the Russian federation. Clim Change 42:219–242
53. Zhai P, Sun A, Ren F, Liu X, Gao B, Zhang Q (1999) Changes of climate extremes in China. Clim Change 42:203–218
54. Türke SM (1996) Spatial and temporal analysis of annual rainfall variations in Turkey. Int J Climatol 16:1057–1076
55. Hess TM, Stephens W, Maryah UM (1995) Rainfall trends in the north east arid zone of Nigeria 1961–1990. Agric For Meteorol 74:87–97
56. Mason SJ (1996) Climatic change over the lowveld of South Africa. Clim Change 32:35–54
57. Heino R, Forland E, Tuomenvirta H, Alexandersson H, Beniston M, Pfister C, Rebetez M, Rosenhagen G, Jones PD, Horton EB, Folland CK, Hulme M, Parker DE, Basnett TA (1999) The use of indices to identify changes in climatic extremes. Clim Change 42:131–149
58. Bootsma A (1994) Long-term (100 years) climate trends for agriculture at selected locations in Canada. Clim Change 26:65–88
59. Smit B, Ludlow L, Brklacich M (1988) Implications of a global climatic warming for agriculture: a review and appraisal. J Environ Q 17:519–527
60. Zhang X, Vincent LA, Hogg WD, Niitsoo A (2000) Temperature and precipitation trends in Canada during the 20th century. Atmos Ocean 38:395–429
61. Gonzalez Hidalgo JC, De Luis M, Raventos J, Sanchez JR (2003) Daily rainfall trend in the valencia region of Spain. Theoret Appl Climatol 75:117–130
62. Gong DY, Shi PJ, Wang JA (2004) Daily precipitation changes in the semi-arid region over northern China. J Arid Environ 59:771–784
63. Del Rio S, Penas A, Fraile R (2005) Analysis of recent climatic variations in castle and leon (Spain). Atmos Res 73:69–85
64. Partal T, Kahya E (2006) Trend analysis in Turkesh precipitation data. Hydrol Process 20:2011–2026
65. Balling RC, Skindlow JA, Philips DA (1990) The impacts of increasing summer mean temperatures on extreme maximum and minimum temperatures in Phoenix, Arizona. J Clim 3:1491–1494
66. Esterling DR, Evans LG, Groisman PY, Karl TR, Kunkel KE, Ambenje P (2000) Observed variability and trends in extreme climate events: a brief review. Bull Am Meteorol Soc 3:417–425
67. Esterling DR, Horton B (1997) Maximum and minimum temperature trends for the globe. Science 227:364–367
68. Nicholls N, Gruza GV, Jouzel J, Karl TR, Ogallo LA, Parker DE (1996) Observed climate variability and change. In: Houghton JT, Filho LGM, Callander BA, Harris N, Kattenberg A, Maskell K (eds) Climate change 1995: the science of climate change. Cambridge University Press, Cambridge, pp 133–192
69. Unkasevic D, Vujovic D, Tosic I (2005) Trends in extreme summer temperatures at Belgrade. Theoret Appl Climatol 82:199–205
70. Mearns LO, Katz RW, Schneider SH (1984) Extreme high temperature events: changes in theri probabilities with changes in mean temperature. J Climatol Appl Meteorol 23:1601–1613
71. Hansen J, Fung I, Lacis A, Rind D, Lebedeff S, Ruedy R, Russel G, Stone P (1988) Global climate change as forecast by goddard institute for space studies three dimensional model. J Geophys Res 93:9341–9354

72. Ghosh S (2011) Impact of climate change and land use change on the flood vulnerability of Brahmaputra basin. Geospatial world forum, 18–21 Jan 2011, Hydarabad, India
73. Jones PD, Moberg A (2003) Hemispheric and large-scale surface air temperature variations: an extensive revision and an update to 2001. J Clim 16:206–223
74. Douville H, Chauvin F, Planton S, Royer J-F, Salas-Mélia D, Tyteca S (2002) Sensitivity of the hydrological cycle to increasing amounts of greenhouse gases and aerosols. Clim Dyn 20:45–68
75. Intergovernmental Panel on Climate Change (IPCC) (2001a) In: McCarthy JJ, Canziani OF, Leary NA, Dokken DJ, White KS (eds) Climate change 2001: impacts, adaptation and vulnerability, contribution of the working group II to the third assessment report of the intergovernmental panel on climate change, Cambridge University Press, Cambridge, UK, p 1032
76. McCarthy JJ, Canziani OF, Leary NA, Dokken DJ, White KS (2001): Climate change 2001: Impacts, adaptation, and vulnerability. Cambridge University Press, p 1032
77. European Environment Agency (EEA) (2008) River floods (CLIM 017)—assessment published Sep 2008. Available online at: http://www.eea.europa.eu/data-and-maps/indicators/river-floods/river-floods-assessment-published-sep-2008. Verified 28 Nov 2011
78. Parry ML, Canziani OF, Palutikof JP, van der Linden PJ, Hansen CD (eds) (2007) Climate change 2007: impacts, adaptation and vulnerability. Contribution of Working Group II to the Fourth Assessment Report of the Intergovernmental Panel on Climate Change. Cambridge University Press, Cambridge
79. Brázdil R, Kotyza O, Dobrovolný P (2006) July 1432 and August 2002–two millennial floods in Bohemia? Hydrol Sci J 51(5):848–863
80. Cyberski J, Grześ M, Gutry-Korycka M, Nachlik E, Kundzewicz Z (2006) History of floods on the river Vistula. Hydrol Sci J 51(5):799–817
81. Kundzewicz ZW (2008) Flood risk and vulnerability in the changing climate, annals of Warsaw university of life sciences—SGGW. Land Reclam 39:21–30
82. Mudelsee M, Börngen M, Tetzlaff G, Grünewald U (2003) No upward trends in the occurence of extreme floods in central Europe. Nature 425:166–169
83. Palmer TN, Räisänen J (2002) Quantifying the risk of extreme seasonal precipitation in a changing climate. Nature 415:512–514
84. Kay AL, Jones RG, Reynard NS (2006) RCM rainfall for UK flood frequency estimation. II. Climate change results. J Hydrol 318:163–172
85. Graham LP, Andre′asson J, Carlsson B (2007) Assessing climate change impacts on hydrology from an ensemble of regional climate models, model scales and linking methods—a case study on the Lule river basin. Clim Change 81:293–307
86. Booij MJ (2005) Impact of climate change on river flooding assessed with different spatial model resolutions. J Hydrol 303:176–198
87. Shabalova M, Van Deursen W, Buishand T (2003) Assessing future discharge of the river Rhine using RCM integrations and a hydrological model. Clim Res 23:233–246
88. Lehner B, Doll P, Alcamo J, Henrichs T, Kaspar F (2006) Estimating the impact of global change on flood and drought risks in Europe: a continental integrated analysis. Clim Change 75:273–299
89. Jain S, Lall U (2000) Magnitude and timing of annual maximum floods: trends and large scale climatic associations for the Blacksmith Fork River, Utah. Water Resour Res 36:3641–3651
90. Sharma KP, Moore B, Vorosmarty CJ (2000) Anthropogenic, climatic and hydrological trend in the Kosi basin. Himalaya, Clim Change 47:141–165
91. Kothyari UC, Singh VP (1996) Rainfall and temperature trends in India. Hydrol Process 10:357–372
92. Meher-Homji VM (1991) Probable impact of deforestation on hydrological processes. Clim Change 19:163–173

93. Malet J-P, Maquaire O, Remaitre A, Thiery Y, Peyron M, Waeckel C (2010) Landslide hazard and risk assessment in the Barcelonnette basin, Mountain risks, research training networks, Causeries de L'e'te'—Barcelonnette, 23 June 2010
94. Buma JT, Dehn M (1996) Development of a method for predicting the impact of climate change on slope stability, Workshop on landslide and flash flood, Barcelonnette—Vaison la Romaine 30 Sept–3 Oct 1996
95. Buma JT, Dehn M (2000) Impact of climate change on landslide in South East France, simulated using different GCM scenarios and downscaling methods for local precipitation. Clim Res 15:69–81
96. Kundzewicz ZW, Schellnhuber H-J (2004) Floods in the IPCC TAR perspective. Nat Hazards 31:111–128
97. Schumann G, Hostache R, Puech C, Hoffmann L, Matgen P, Pappenberger F, Pfister L (2007) High-resolution 3-D flood information from radar imagery for flood hazard management. IEEE Trans Geosci Remote Sens 45(6):1715–1725
98. Van Stokkom HTC, Smits AJM, Leuven RSEW (2005) Flood defence in the Netherlands: a new era, a new approach. Water Int 30:76–87
99. Yuan F, Sawaya K, Loeffelholz B, Bauer M (2005) Land cover classification and change analysis of the twin cities (Minnesota) metropolitan area by multitemporal landsat remote sensing. Remote Sens Environ 98:317–328
100. Brockerhof M (2000) An urbanizing world. Population Bull 55(3):1–44
101. Orr HG, Carling PA (2006) Hydro-climatic and land use changes in the river Lune catchment, North West England, implications for catchment management. River Res Appl 22(2):239–255
102. Rosso R, Rulli MC (2002). An integrated simulation method for flash-flood risk assessment: 2. Effects of changes in land-use under a historical perspective. Hydrol Earth Syst Sci 6(2):285–294
103. Tollan A (2002) Land-use change and floods: what do we need most, research or management? Water Sci Technol 45(8):183–190
104. Patric JH, Reinhart GK (1971) Hydrologic effects of deforesting two mountain watersheds in west Virginia. Water Resour Res 7(5):1182–1188
105. Bannister EN (1979) Impact of road networks on south eastern Michigan lakeshore drainage. Water Resour 15:1515–1520
106. Costa MH, Botta A, Cardille JA (2003) Effects of large-scale changes in land cover on the discharge of the Tocantins River. Southeast Amazonia, J Hydrol 283:206–217
107. Brath A, Montanari A (2000) Effects of the spatial variability of soil infiltration capacity in distributed rainfall runoff modelling. Hydrol Process 14:2779–2794
108. Naef F, Scherrer S, Weiler M (2002) A process based assessment of the potential to reduce flood runoff by land use change. J Hydrol 267(1–2):74–79
109. Bosch JM, Hewlett JD (1982) A review of catchment experiments to determine the effect of vegetation changes on water yield and evapotranspiration. J Hydrol 55:3–23
110. Beighley RE, Moglen GE (2002) Trend assessment in rainfall runoff behaviour in urbanizing watersheds. J Hydrol Eng 7:27–34
111. Luft G, Morgenschweis G, Vogelbacher A (1982) The effects of large-scale terracing on hydrological processes, Presentation of first results of a study comparing two small experimental basins on the Kaiserstuhl-Mountain. Proceedings of symposium on hydrological research basins, Sonderheft Landeshydrologie, Bern, pp 543–553
112. Ranzi R, Bochicchio M, Bacchi B (2002) Effects on floods of recent afforestation and urbanisation in the Mella River (Italian Alps). Hydrol Earth Syst Sci 6:239–253
113. Peck AJ, Williamson DR (1987) Effects of forest clearing on groundwater. J Hydrol 94:47–65
114. Armando B, Monanari A, Moretti G (2006) Assessing the effect on flood frequency of land use change via hydrological simulation (with uncertainty). J Hydrol 324:141–153
115. Sullivan A, Ternan JL, Williams AG (2004) Land use change and hydrological response in the camel catchment. Cornwall Appl Geogr 24(2):119–137

116. De Roo APJ, Schmuck G, Perdigao V, Thielen J (2003) The influence of historic land use changes and future planned land use scenarios on floods in the oder catchment. Phys Chem Earth (B) 28(33–36):1291–1300
117. Reynard NS, Prudhomme C, Crooks SM (2001) The flood characteristics of large UK rivers: potential effects of changing climate and land use. Clim Change 48:343–359
118. McMillan HK (2002) The estimation of discharges in ungauged sub-catchments of the River Eden in Cumbria, Unpublished Master Thesis, Lancaster University
119. Bultot F, Dupriez GL, Gellens D (1990) Simulation of land-use changes and impacts on the water balance, a case study for Belgium. J Hydrol 114:327–348
120. Eiker E, Davis DW, Goldman DM (2000) Application of risk-based analysis to planning reservoir and levee flood damage reduction systems. Us Army Corps of Engineers Hydrologic Engineering Center (HEC), USA, pp 1–30
121. Remaitre A, Malet J-P (2010) The effectiveness of torrent check dams to control channel instability: example of debris-flow events in clay shales. In: Garcia CC, Lenzi MA (eds) Check dams, morphological adjustments and erosion control in torrential streams. Nova Science Publishers Inc., New York, pp 211–237
122. Bravard J-P (1986) Le Rh6ne, du Leman a Lyon. p 450, La Manufacture, Lyon. Peiry 1988 (In French)
123. Bravard J-P (1989) La metamorphose des rivieres des Alpes francaises a la fin du moyen-age et a l'epoque moderne. Bull Soc Geog Liege 25:145–157 (In French)
124. Schumm SA (1977) The fluvial system. Wiley, New York 338
125. Hey RD, Thorne CR (1986) Stable channels with mobile gravel-bed rivers. J Hydraul Eng 8:671–689
126. Gonde R (1981) Les amenagements du lit de la Durance et l'evolution du milieu naturel, p 65. Report of Ministere de l'Environnement et du Cadre de Vie, Paris (In French)
127. Piegay H, Bravard JP, Dupont P (1994) Les ripisylves et les crues dans la France du sud-est: de l'histoire a la gestion contemporaine. 23emes Journees de I'Hydraulique: Crues et inondations (ed. by Societ6 Hydrotechnique de France), pp 277–289. Nimes, France (In French)
128. Penning-Rowsell EC, Johnson C, Tunstall S, Tapsell S, Morris J, Chatterton J, Green C (2005) The benefits of flood and coastal risk management: a manual of assessment techniques. Middlesex University Press, London
129. Bhattacharya N (2010) Flood risk assessment in Barcelonnette, France, M.Sc. thesis, International Institute for Geo-information Science and Earth Observation (ITC), Twente University, Enschede, The Netherlands
130. Alkema D (2007) Simulating floods, on the application of a 2D-hydraulic model for flood hazard and risk assessment, Ph.D. Dissertation, International Institute for Geo-information Science and Earth Observation (ITC), Twente University, Enschede, The Netherlands. ISBN 978 90 6164 263 3
131. Apel H, Thieken A, Merz B, Blöschl G (2006) A probabilistic modelling system for assessing flood risks. Nat Hazards 38:79–100
132. Jain V, Sinha R (2006) Evaluation of geomorphic control on flood hazard through geomorphic instantaneous unit hydrograph. Current science association
133. Boughton WC, Droop O (2003) Continuous simulation for design flood estimation—a review. Environ Model Softw 18(4):309–318
134. Plate EJ (2007) Flood risk management for setting priorities in decision making, Springer. Vasiliev OF, van Gelder PHAJM, Plate EJ, Bolgov MV (eds) Extreme hydrological events: new concepts for security. pp 21–44
135. World Meteorological Organization (WMO) (1981c) Meteorological and hydrological aspects of siting and operation of nuclear power plants, vol II Hydrological aspects, technical note no. 170, (WMONo. 550), Geneva, Switzerland, p 125
136. Klein Tank AMG, Wijngaard JB, Konnen GP, Bohm R, Demaree G, Gocheva A, Mileta M, Pashiardis S, Hejkrlik L, Kern-Hansen C, Heino R, Bessemoulin P, Muller-Westermeier G,

Tzanakou M, Szalai S, Palsdottir T, Fitzgerald D, Rubin S, Capaldo M, Maugeri M, Leitass A, Bukantis A, Aberfeld R, VanEngelen AFV, Forland E, Mietus M, Coelho F, Mares C, Razuvaev V, Nieplova E, Cegnar T, López JA, Dahlstrom B, Moberg A, Kirchhofer W, Ceylan A, Pachaliuk O, Alexander LV, Petrovic P (2002) Daily dataset of 20th-century surface air temperature and precipitation series for the European climate assessment. Int J Climatol 22:1441–1453

137. Klein Tank AMG, Können GP (2003) Trends in indices of daily temperature and precipitation extremes in Europe. J Clim 16:3665–3680
138. Frich P, Alexander LV, Della-Marta P, Gleason B, Haylock M, Tankand AMGK, Peterson T (2002) Observed coherent changes in climatic extremes during the second half of the twentieth century. Clim Res 19:193–212
139. Alexander LV, Zhang X, Peterson TC, Caesar J, Gleason B, Klein Tank AMG, Haylock M, Collins D, Trewin D, Rahimzadeh F, Tagipour A, Rupa Kumar K, Revadekar J, Griffiths G, Vincent L, Stephenson DB, Burn J, Aguilar E, Brunet M, Taylor M, New M, Zhai P, Rusticucci M, Vázquez Aguirre JL (2006) Global observed changes in daily climate extremes of temperature and precipitation. J Geophys Res 111:1–22
140. Norrant C, Douguédroit A (2006) Monthly and daily precipitation trends in the mediterranean. Theor Appl Climatol 83:89–106
141. Piegay H, Salvador P-G (1997) Contemporary floodplain forest evolution along the middle Ubaye river. Southern Alps, France, Global Ecology and Biogeography Letters 6:397–406
142. Demontzey P (1882) Traite pratique de reboisement et du gazonnement des montagnes. J. Rothschild, Paris (In French), p 528
143. Lecarpentier C (1963) La crue de Juin 1957 en Ubaye et ses conséquences morphodynamiques. Ph.D. Dissertation, Faculty of Geography, Strasbourg University, p 319 (In French)
144. Sclafert T (1933) A propos du d6boisement des Alpes du Sud. Ann Geogr 42:266–277 (In French)
145. Salvador PG (1991) Le theme de la metamorphosefluviale dans lesplaines alluviales du Rh6ne et de l'Isere (Bassin de Malville et Ombilic de Moirans, Bas-Dauphine), p 529. Th&se de geographie, Universite Lyon III, Lyon (In French)

Chapter 3
Methodology

3.1 Introduction

This chapter is first constituted of the conceptual and methodological approaches leading the research on comparisons between digital elevation model (DEM) quality control modeling and interpolation techniques. Second, it intends to explain the input data preparation procedures for hydrodynamic modeling. Then in the next step, the calibration and modeling approaches for flood hazard assessment, the basic theory for hydraulic and hydrologic models (HEC-RAS and SOBEK), are described. Following this, the statistical methods applied to randomness, homogeneity, and trend detection in hydrometeorological variables are described. Finally, in the last part of this chapter, the approaches used to analyze the land use changes from the past until the present are posed and presented.

Model analysis such as land use change detection and DEM quality control modeling was mostly done using ESRI© software ArcGIS 9.3.1, based on different extensions and tools such as: Spatial Analyst, 3D Analyst, and Geostatistical Analysis, Arc Hydro Tools, HEC-GeoRAS, Hawthstools-Sampeling Tools coupled with SAM.

For hydrological simulation, HEC-RAS (1D), developed by the US Department of Defense Army Corps of Engineers, coupled with the SOBEK (1D–2D) model, developed by Delft Hydraulics (The Netherlands), were used to analyze flood hazard scenarios and flow characteristics over time. ArcGIS and ILWIS softwares were used in the pre- and post-modeling phases of HEC-RAS and SOBEK for data preparation, visualization, and analysis.

The calculation of trends in climatic parameters was made with SPSS software. Land use change detection was analyzed in ArcGIS. The scale of all GIS layers was fixed in ArcGIS using the "Raster Storage Setting". From different pyramid resampling techniques, the "Nearest Neighbor Method", which is based on performing a nearest neighbor assignment, was selected because it is a fast interpolation

A. Ramesh, *Response of Flood Events to Land Use and Climate Change*,
Springer Theses, DOI: 10.1007/978-94-007-5527-7_3,

method useful for land use classification with a more smoothing capacity. Then, all GIS layers were projected onto the following projection system:

NTF_Lambert_Zone_III
Projection: Lambert_Conformal_Conic
False_Easting: 600000.00000000
False_Northing: 200000.00000000
Central_Meridian: 2.33722917
Standard_Parallel_1: 43.19929139
Standard_Parallel_2: 44.99609389
Scale_Factor: 1.00000000
Latitude_Of_Origin: 44.10000000
Linear Unit: Meter
Geographic Coordinate System: GCS_NTF
Datum: D_NTF
Prime Meridian: Greenwich
Angular Unit: Degree

3.2 DEM Quality Control Modeling

This part was carried out to analyze the effect of statistically improved DEM on the outputs of the flood inundation model in comparison with the existing DEM. According to Vaze [1], one of the most commonly used and widely available basic spatial information models, especially in hydrodynamic simulation, is the DEM. Alkema [2] also realized in the related research that within hydraulic modeling, the output is affected to a large extent by model input parameters, and a reliable input data set is required to accomplish successful modeling [3]. Therefore, in preparing the input data, a major emphasis has to be placed to increase the quality of the DEM as much as possible.

The DEM applied in this research, with 5 and 10 m resolutions, was derived from Radar, which was retrieved from the OMIV Coordinator, OSUG-LGIT of France. The DEM was controlled using a variety of interpolation methods. All the interpolation methods were available in the Geostatistical Analyst extension of ArcGIS and use observed points as input data. The selected interpolation methods were commonly used and suggested by many investigators

3.2.1 *Interpolation Techniques*

The quality of a derived DEM can vary greatly depending on the data source and the interpolation technique which is applied. Burrough and McDonnell [4] have defined the interpolation method as a technique to construct and predict the new data points within the range of known data points. They also mentioned that in the fields of engineering and science there are often a number of data points obtained by different methods such as sampling or experimentation. These methods try to construct a function which is closely fitted to all those data points. This constructing of a function between the data points is called curve fitting or regression analysis. Therefore, interpolation is a specific case of curve fitting, in which the function must go exactly through the data points. According to Alkema and Rahman [5], the interpolation methods are considered to be designed to create statistically correct surfaces.

There are two main interpolation techniques:

1. Deterministic: This method is based on mathematical equations to predict the new data points based on the assumption that the interpolating surface should be influenced mostly by the nearby points and less by the more distant points.
2. Geostatistic: This method applies both mathematical and statistical approaches to predict the new data points at unknown locations. This method is based on the assumption that the interpolating surface should be influenced mostly by the nearby points, and less by the more distant points. Nevertheless, it is also based on the spatial autocorrelation among data points [4, 6]. Geostatistical techniques quantify the spatial autocorrelation among measured points and account for the spatial configuration of the sample points around the prediction location [7].

There are typical methods used to produce many of the DEMs in use today or to correct errors. In this study, the main aim was to compare the quality of different interpolation techniques available to derive DEMs from the point data. The interpolation techniques were used also to determine DEM errors. The interpolation methods compared were some common interpolation algorithms based on the inverse distance weighted (IDW), Topo to Raster, and Spline as deterministic methods. In Geostatistical analysis, Kriging (Spherical, Circular, Exponential, Gaussian, and Stable) was selected. Figure 3.1 shows the process of DEM quality control modeling based on different interpolation techniques.

Spline and Inverse Distance Weighting were chosen because they are commonly used now and were also used in earlier work, and they have been recommended by many investigators [8, 9]. The Topo to Raster method was selected because it automatically removes spurious sinks, has advantages for faster computation, lowers "roughness penalty" [10], and has a drainage enforcement algorithm for preserving the natural sinks in the area [11]. The Kriging method was found to produce the best results in general tests of interpolators. All methods were implemented in the ArcGIS environment and were thus easy to access. The interpolation methods applied in this research have different characteristics as the following table outlines (Table 3.1).

The detailed interpretation with related equations for all the above-mentioned techniques is described in the next section.

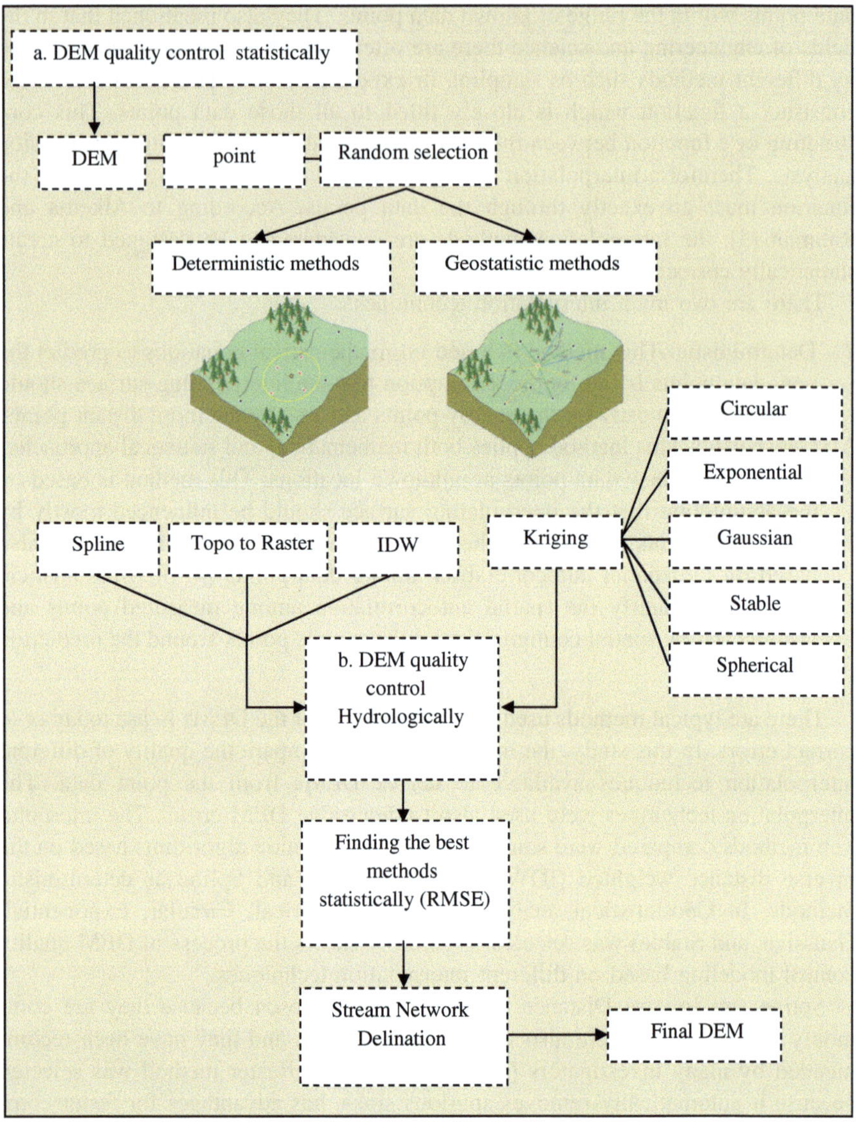

Fig. 3.1 DEM quality control procedures

Table 3.1 Different interpolation characteristics (Modified information based on [1, 14, 19, 20, 78, 139]

Method	Speed	Comments
Spline	Fast	Available in ArcGIS; estimates values using a mathematical function that minimizes overall surface curvature.
IDW	Fast	Available in ArcGIS; tends to generate bull's eye patterns; simple and effective with dense data; no extrapolation.
Topo to Raster	Slow/Medium	Available in ArcGIS; interpolates a hydrologically "correct" grid from a set of point, based on procedures developed by Hutchinson [19, 20, 78].
Kriging	Slow	Available in ArcGIS; very flexible range of methods based on modeling variogram; can provide extrapolation and prediction error estimates; speed not substantially affected by increasing the number of data points.

Spline

The Spline method applies mathematical functions minimized on the overall surface curvature to estimate the unknown values. The Spline results represent a smooth surface passed exactly through the input points while minimizing the total curvature of the surface [12, 13; 14]. The algorithm used for the Spline method applies the following equations:

$$S(x,y) = T(x,y) + \sum_{j=1}^{N} \lambda_j R(r_j) \tag{3.1}$$

where:

j is 1, 2,..., N; N is the number of points; λ_j are coefficients found by the solution of a system of linear equations; r_j is the distance from the point (x, y) to the jth point.

For the regularized option the formula is as follows:

$$T(x,y) = a_i + a_i x + a_3 y \tag{3.2}$$

where:

a_i are coefficients found by the solution of a system of linear equations. And,

$$R(r) = \frac{1}{2\pi}\left\{\frac{r^2}{4}\left[ln\left(\frac{r}{2\tau}\right) + c - 1\right] + \tau^2\left[K_0\left(\frac{r}{\tau}\right) + c + ln\left(\frac{r}{2\pi}\right)\right]\right\} \tag{3.3}$$

where:

r is the distance between the predicted point and the sample data points; c is defined as a constant which is equal to 0.577215; τ^2 is the Weight parameter; K_0 is the modified function.

Inverse Distance Weighted

One of the most commonly used interpolation techniques of scatter points is IDW. The IDW method is based on the assumption that the interpolating surface should be influenced mostly by the nearby points and less by the more distant points (http://webhelp.esri.com). IDW deterministic interpolation technique applies the idea that influence decreases with increasing the distance from particular points.

Burroughs and McDonnell [4], Philip and Watson [15], Watson and Philip [15] realized that "the value at un-sampled points is a distance-weighted average of values at sampled locations within a defined neighborhood surrounding the un-sampled point". In this sense, Johnston [6] found that the points closer to the prediction location will have more influence on the predicted value than points located farther away [6].

To fill an elevation matrix, IDW technique uses weighted averaging techniques. This means that the interpolated values of a cell are determined from the values of nearby data points, taking into account the distance of the cell from those input points [17]. The equation of IDW for interpolating the elevation is shown as:

$$\widehat{Z}(s_0) = \sum_{i=1}^{N} \lambda_i . z(s_i) \tag{3.4}$$

where:

$\widehat{Z}(s_0)$ is the predicted value at the unsampled points S_0; N is the number of measured sample points within the neighborhood defined for S_0; λ_i are the distance-dependent weights associated with each sample point; and z λ_i is the observed value at location S_i. Weights are calculated using:

$$\lambda_i = \frac{d_{i0}^{-p}}{\sum_{i=1}^{N} d_{i0}^{-p}} \tag{3.5}$$

$$\sum_{i=1}^{N} \lambda_i = 1 \tag{3.6}$$

where:

d_{i0} is the distance between the prediction values at unknown location S_0 and the measured values at known location S_i; P is the power parameter which is applied to define the rate of reduction of the weights as the distance increases [4, 6, 18].

Topo to Raster

Topo to Raster is a specific technique designed to convert the vector surface data, in the form of contours, into a consistent DEM format hydrologically. It is based on interpolation models and the Australian national university digital elevation model (ANUDEM) program developed by Hutchinson [19, 20], specifically designed for the creation of hydrologically correct DEMs which are provided within ArcGIS.

The interpolation procedure has been designed to take advantage of the types of input data commonly available and the known characteristics of elevation surfaces. This method uses an iterative finite difference interpolation technique.

It is optimized to have the computational efficiency of local interpolation methods, such as IDW interpolation, without losing the surface continuity of global interpolation methods, such as Kriging and Spline. It is essentially a discretized thin plate of Spline technique [19–21], for which the roughness penalty has been modified to allow the fitted DEM to follow abrupt changes in terrain, such as streams and ridges. It is also the only ArcGIS interpolator specifically designed to work intelligently with contour inputs.
The roughness penalty is defined in function of the first- and second-order partial derivatives of the interpolation function *f* by the following equations:

$$J_1(f) = \int \left(f_x^2 + f_y^2\right) \mathrm{d}x\mathrm{d}y \tag{3.7}$$

$$J_2(f) = \int \left(f_{xx}^2 + 2f_{xy}^2 + f_{yy}^2\right) \mathrm{d}x\mathrm{d}y \tag{3.8}$$

where:

minimizing J_1 in its discretized form conduces to discretized minimum potential interpolation, while minimizing J_2 conduces to the minimum curvature interpolation of thin-plate splines in their discrete form. If only J_2 was minimized, the resulting surface would be unrealistically smooth, whereas minimizing J_1 gives rise to sharper local maxima and minima at data points as the grid resolution becomes finer. Hutchinson [20] suggested the following compromise between J_1 and J_2 to define the roughness penalty in this algorithm:

$$J(f) = 0.5 \times h^{-2} J_1(f) + J_2(f) \tag{3.9}$$

where:

h is the cell resolution; this form of roughness penalty maintains trends beyond data points as in minimum curvature interpolation, allows the definition of sharp changes in slope occurring at ridges, and identifies points at rivers as sinks.

Kriging

Autocorrelation in spatial phenomena is a defining concept of Geostatistic interpolation methods. On the other hand, the data points which are located close together tend to be more similar than the data points which are located far from each other. Therefore, in this case, the Kriging method is an interpolation technique which is similar to the IDW method. The only difference between Kriging and the deterministic method such as IDW is that the procedure to construct the new values is not only based on the distance between the samples and the prediction location, but also on the estimate according to the spatial arrangement of the sampled points. This means that the measurement is based on spatial autocorrelation between these points [6]. Kriging is represented by the Spherical, Circular, Exponential, Gaussian, and Stable methods. With these options, Kriging uses the mathematical function specified with the argument to fit a line or curve to the semivariance data in the semivariogram. Ordinary Kriging assumes that the variation in z-values is free of any structural component. These five models are provided to ensure the necessary conditions of the variogram model are satisfied, which is not always possible with interactive, manual variogram fitting. These methods and conditions are discussed in McBratney and Webster [22]. They described the methodology process as:

1. First, the variance is calculated based on the average variance of all point pairs within each interval of the cell size.
2. Then, the variogram is fitted to the variance points using the Levenberg–Marquardt Method [23] of a nonlinear least squares approximation. A minimum of three points of data are required for the fit. In mathematics and computing, the Levenberg–Marquardt algorithm (LMA) provides a numerical solution to the problem of minimizing a function, generally nonlinear, over a space of parameters of the function. These minimization problems arise especially in least squares curve fitting and nonlinear programming [24].
3. Next, by increasing the cell size, the number of sample points per cell size interval will increase, thereby providing enough data points to estimate the semivariogram.
4. Once the semivariogram is estimated, a smaller cell size can be used in creating the actual output grid.
5. After the construction of the empirical semivariogram, a model is fitted to the plotted values using a defined function.

In general, Kriging goes through a two-step process:

- Variograms and covariance functions are created to estimate the statistical dependence (spatial autocorrelation) values, which depend on the model of autocorrelation (fitting a model).
- Prediction of unknown values. Therefore, The Kriging model is defined as:

$$Z(s) = \mu + \varepsilon(s) \tag{3.10}$$

where:

$Z(s)$ is the value at an unknown location of s; μ is a constant mean for the data; and $\varepsilon(s)$ is defined as dependent random errors.

In Kriging, to predict the values at unknown points, the following equation would be applied:

$$\widehat{Z}(s_0) = \sum_{i=1}^{N} \lambda_i . z(s_i) \tag{3.11}$$

where:

$\widehat{Z}(s_0)$ is the predicted value at the unsampled location S_0; N is the number of measured sample data points within the neighborhood defined for S_0; λ_i are the weights associated with each sample point; and $z\ (s_i)$ is defined as the observed value at location S_i.

In Kriging, the average difference between the predicted and the measured value should be zero. To achieve this, the value of the following equation has to be minimized:

$$\left(\widehat{Z}(s_0) = \sum_{i=1}^{N} \lambda_i . z(s_i)\right)^2 \tag{3.12}$$

The minimization of the above formula generates the Kriging equations:

$$\boldsymbol{\Gamma} \times \lambda = g \tag{3.13}$$

where:

the variance–covariance matrix is defined as $\boldsymbol{\Gamma}$ which is calculated from the data points, multiplied by the column vector containing the weights being calculated, which is defined as λ equal g which is a column containing the fitted semivariance for the predicted location.

Fitting a Model to the Empirical Semivariogram

The next step in Kriging procedures is to fit a model to the data points by fitting the empirical semivariogram. In the previous section, it was explained how the empirical semivariogram provides information on the spatial autocorrelation of datasets. However, it does not provide information for all possible directions and distances. For this reason, and to ensure that Kriging predictions have positive Kriging variances, it is necessary to fit a model to the empirical semivariogram. Fitting a model or variography with the structural analysis is computed as:

$$\text{Semivariogram}(\text{distance}h) = 0.5 \times \text{average}\left[(\text{valueatlocation}i - \text{valueatlocation}j)^2\right]$$

The equation represents the difference squared between the values of the paired locations. Figure 3.2 shows the pairing of one point (the dark blue point) with all other measured locations. This process continues for each measured point. Each pair of locations has a unique distance, and there are often many pairs of points.

To plot all pairs quickly becomes unmanageable. Instead of plotting each pair, the pairs are grouped into lag bins; for example, computing the average

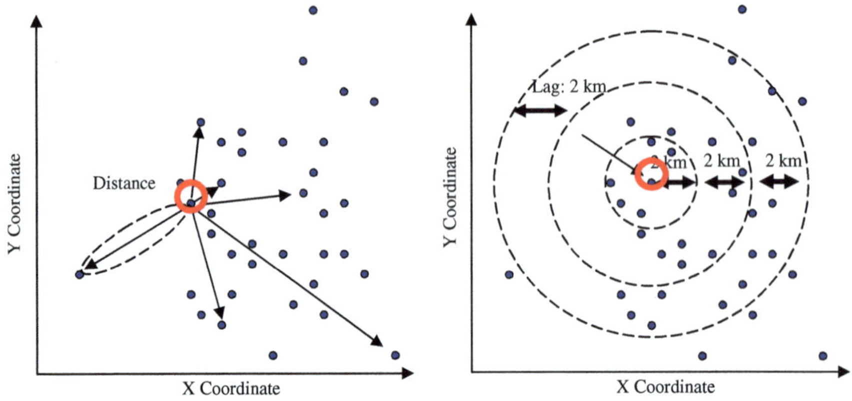

Fig. 3.2 Pairing of one point (*inside of red circular*) with all other measured locations [138]

Fig. 3.3 Hypothetical example of an empirical semivariogram and its fitted model; the distance where the model first flattens out is known as the range; the value at which the model attains the range is called the sill; the value at which the model intercepts the *y*-axis is called the nugget [6]

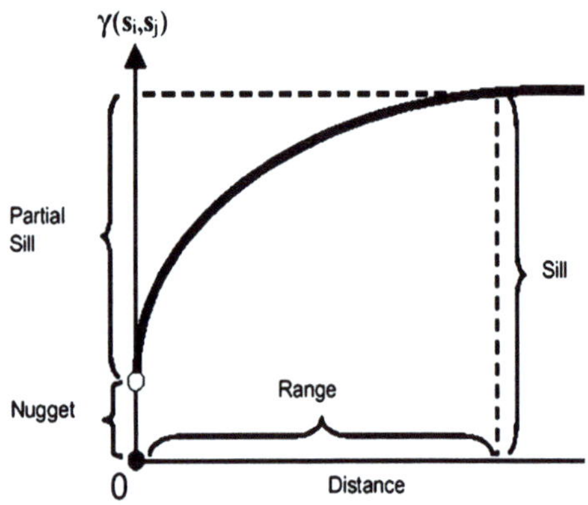

semivariance for all pairs of points that are greater than 40 m apart but less than 50 m. The empirical semivariogram is defined as a graph of the averaged semivariogram values on the *y*-axis and the lag or distance on the *x*-axis (Fig. 3.3).

According to Fig. 3.3, Johnston [6] has defined the Lag, Range, Sill, and Nugget as follows:

- Lag is a distance interval in which whole existing pairs are averaged (similarity or dissimilarity) to simplify the calculation;
- Range is separation between point pairs at which the sill is reached; distance at which there is no evidence of spatial dependence;
- Sill is maximum semivariance; represents variability in the absence of spatial dependence;
- Nugget is semivariance as the separation approaches zero; represents variability at a point that cannot be explained by spatial structure.

Spatial autocorrelation quantifies a basic principle of geography that the values which are located closer are more similar than the values that are located farther

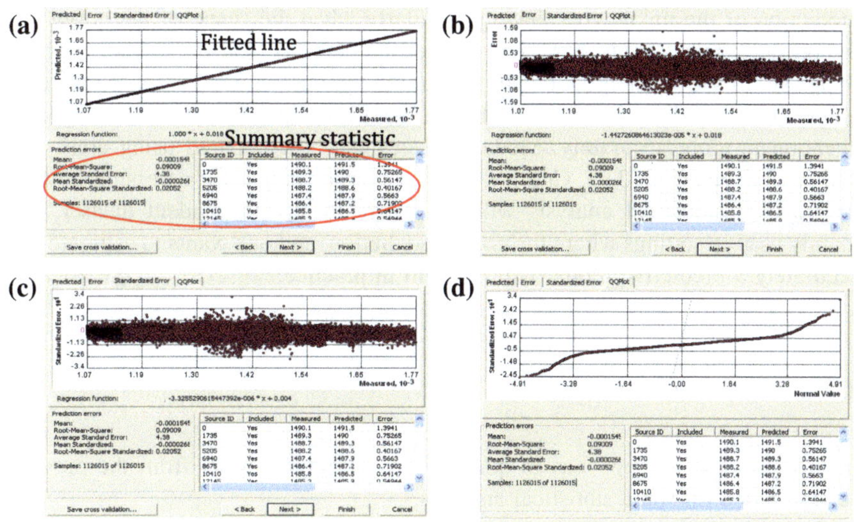

Fig. 3.4 Cross validation statistic; **a** Cross validation between measured and predicted values, **b** Estimated error, **c** Standardized error, **d** Q–Q plot

apart. Thus, pairs of locations that are closer (far left on the *x*-axis of the semivariogram cloud) should have more similar values (low on the *y*-axis of the semivariogram cloud). As pairs of locations become farther apart (moving to the right on the *x*-axis of the semivariogram cloud), they should become more dissimilar and have a higher squared difference (move up on the *y*-axis of the semivariogram).

The implementation of Kriging begins with the calculation of the empirical semivarigram. To do this, the empirical semivariance between each pair of data points is calculated as:

$$\gamma_{ij} = 0.5 \times \left[\left(\left(z(s_i) - z(s_j) \right)^2 \right] \tag{3.14}$$

With large datasets the calculation of all possible combinations of the points is not feasible. Therefore, data points are grouped based on their distance and direction from one another in a process called binning. To construct the empirical semivariogram graph, the average semivariogram value for all the pairs within each bin is plotted against the average distance [7].

To fit a model to the empirical semivariogram, some functions such as Spherical, Exponential, Stable, Gaussian, etc., should be selected, which rise at first and then level off for larger distances beyond a certain range. From the model, there are deviations of the points on the empirical semivariogram. Some points are located above the model curve and some are located below. But if we add the distance of each point, those located above the line, with those located below the line, the two values should be similar. As mentioned before, there are different functions and models. Selecting the best model is important and it influences the prediction of the unknown values, particularly when the shape of the curve near the origin differs significantly. The steeper the curve is near the origin, the more influence the closest neighbors will have on the prediction [7, 25]. As a result, the output surface will be less smooth. Each model is designed to fit different types of phenomena more accurately. Model surfaces can be compared using cross validation statistics (Fig. 3.4). Cross validation is a technique for assessing how the results of a statistical analysis will generalize to an independent dataset. It is mainly used in settings where the goal is prediction, and one wants to estimate how accurately a predictive model will perform in practice.

3.2.2 Statistical Analysis

In this part, the attempt was to quantify the achieved results from the previous section based on some statistical analysis.

Root Mean Square: The root mean square error (RMSE) represents the difference between the measured control points data and the predicted control point locations calculated by the transformation process.

It is expressed as:

$$RMSE_{\text{elev}} = \sqrt{\frac{\sum_{i=1}^{N} (x_i - x_j)^2}{N - 1}} \tag{3.15}$$

where:

x_i is the predicted elevation and x_j is the observed elevation. N is the number of sample points.

Although such a measure of the accuracy of elevations in a DEM is clearly useful and sensible, there are still a number of limitations with RMSE to evaluate the DEM quality as several investigators have already addressed [8, 26–28]:

1. RMSE is a global measure and it does not refer to the spatial characteristics of the error. However, it would be expected that the error in whole DEM varies. Many investigations such as Gong et al. [29], Chang and Tsai [30], Carter [31], Bolstad and Stowe [32] have addressed that the error is strongly linked to the nature of the terrain surface, given the high spatial autocorrelation in elevation data. On the other hand, the error is likely to be highly spatially autocorrelated. Additionally, other researches, such as Albani and Klinkenberg [33], Brown and Bara [34], Guth [35], Wise [7] and Wood and Fisher [36], emphasized that many DEM creation methods produce distinctive artifacts that have a strong spatial signature.
2. Another limitation of RMSE was found by Florinsky [37] and Wise [28]. They mentioned that small errors in predicted elevation data points can produce a large range of errors in derived values.
3. The last limitation of RMSE is perhaps related to the criteria that RMSE is only based on a small sample of known data elevation points [38].

Characteristic of elevation error: To find possible errors, the contour lines for each modeled DEM were compared with the contour lines of original DEM.

Q–Q plot: In statistics, a Q–Q plot ("Q" stands for quantile) is a probability plot, which is a graphical method for comparing two probability distributions by plotting their quantiles against each other [39].

Descriptive statistic between original, estimated, and residuals values: In this part, the Moran's I analysis [40] was applied to analyze the spatial autocorrelation analysis. Moran's I is a measure of spatial autocorrelation developed by Patrick A. P. Moran. Spatial autocorrelation is characterized by a correlation in a signal among nearby locations in space [40]. This analysis explores the spatial patterns in the model especially in model residuals. Spatial Analysis in Macroecology (SAM) was applied to explore spatial patterns in the model [41].

3.2.3 Hydrological Analysis

Different methods have been proposed and suggested by many researchers to address the problem of drainage analysis. These methods range from simple DEM smoothing to flood direction assignment (Dutta et al. 2001; [42–46]). Hutchinson et al. [20, 47] developed one of the most satisfactory methods which assigns to drainage network direction. The innovation behind this method is a drainage enforcement algorithm to improve DEM hydrologically. The method is adopted in an ArcGIS environment, which automatically removes spurious sinks or pits from a fitted grid (Dutta et al. 2001; [20, 47]). To delineate watersheds and stream networks the process developed by Hutchinson [20, 47] is defined as follows:

(a) construct the stream network by selecting a drainage area threshold; (b) divide the stream network into individual stream links; (c) find the outlet cell at the lower end of each link; (d) delineate the watershed for each of these outlet cells. This method works well in hill slopes, but it may have some flaws in flat areas [20, 47, 46].

To test the quality of DEM hydrologically, the best DEMs from the previous step (applying interpolation methods) were selected to create the stream network for each surface using the Arc Hydro tools as an extension of ArcGIS. The modeled drainage network derived from each DEM from the previous sector was compared with the observed stream network. The comparison was based on qualitative evaluation of the level of agreement between the modeled and observed drainage network.

The Arc Hydro tool was applied to derive several data sets which describe the drainage networks of the catchment. Raster analysis is performed to generate data on DEM reconditioning, fill sinks, flow direction, flow accumulation, stream definition, stream segmentation, catchment delineation, and watershed delineation. A short definition for each data set is defined in Table 3.2.

Table 3.2 Definition and description of several data sets to create drainage networks (modified and summarized based on the Arc Hydro tools manual)

Data set	Definition & objective
DEM reconditioning	To modify the DEM based on observed drainage networks
Fill sink	To fill the sinks in a grid
Flow direction	To calculate the direction of flow for each grid
Flow accumulation	To analyze the flow accumulation grid that contains the accumulated number of cells located in the upstream of a selected cell
Stream definition	To define a stream grid value as "1" for all the cells in the input flow accumulation grid that have a value greater than the given threshold. The rest of cells in the stream grid contain no data or zero
Stream segmentation	To define a grid of stream segments with unique identification
Catchment grid delineation	To create a grid code indicating to which catchment the cell belongs
Catchment polygon processing	To create the raster and the vector data stored in a feature dataset named "Layers"
Drainage line processing	To convert the input stream link grid into a drainage line feature class

Table 3.3 Error Matrix resulting from the comparison between modeled and observed drainage networks (modified table according to Peralvo [7])

	Modeled	
Observed	River	No River
River	[a]Correct river	[b]Omission error
No River	[c]Commission error	[d]Correct no river

[a] Representing agreement in river in both modeled and observed is "correct river"
[b] Representing river in observed not defined as such in the modeled river showing "omission error"
[c] Representing river in modeled not present in observed indicate "commission error"
[d] Agreement in non river in modeled and observed river indicate "correct no river"

The results of modeled and observed drainage networks derived from all interpolation methods were estimated visually and tallied into an error matrix (Table 3.3).

According to Table 3.3, representing agreement in river in both modeled and observed is "correct river"; representing river in observed not defined as such in the modeled river showing "omission error"; representing river in modeled not present in observed indicate "commission error", and agreement in nonriver in modeled and observed river indicate "correct no river". Regeneration of DEM, visual interpretation, and hill shade analysis was done to check any unnecessary pits or sinks in the surface.

Additionally, Arc Hydro tools were applied to generate drainage networks correctly and to perform pour point tests to check any unnatural variations within the train.

3.3 Adding Feature Surface to DEM

In flood modeling, the DEM must contain all relevant surface features that affect the flow distribution of water, such as dykes. For this purpose, it was necessary to integrate the man-made terrain in the DEM, including roads, dykes, and buildings, and their respective values were obtained from the fieldwork and previous databases and research in the area. Two separate DEMs were produced: one for the 1957 flood event before reconstruction and the height increase of the dykes, and one after the 1957 flood event when the government decided to increase the height of the dyke by about 0.5 m. Road generalization was made in the form of primary, secondary, service roads, and tracks, and heights of 0.5, 0.3, 0.2, and 1 m were assigned to different kinds of roads respectively, based on the literature review [11]. The building also added to DEM as solid blocks hinder the flow water as it surpassed the building. Due to limited data, heights of approximately 3 m were defined for all buildings.

3.4 RAS: Geometric Data

Both SOBEK and HEC-RAS models applied in the hydrodynamic simulation sector need geometric data surface, roughness values, cross sections, hydrometric data, and DEMs. To prepare the geometric input data, it was necessary to apply the DEM of the river system in a grid format. The HEC-GeoRAS model was applied to prepare all input data before starting flood simulation in both modeling. The process of importing data into HEC-RAS model was done in HEC-GeoRAS models as a set of ArcGIS tools. This extension allows users to create an HEC-RAS import file which consists of geometric data such as banks, dykes, river, and flow path from an existing DEM and complementary data sets.

The process to import data into SOBEK was twofold. First, the input data for 1D features such as rivers or channels had to be imported and, second, the 2D features such as overland flow were added. The 1D features were defined using 1D nodes within the flow model. These nodes were connected as a network using straight line connectors called "reaches". The details are explained in sector (Sect. 3.9.4).

3.5 Creating Cross Section Using HEC-GeoRAS Model

Cross section data were applied to determine the conveyance and storage of flow to the channel and floodplain areas. Cross section must be obtained at sufficient intervals and lengths to define the flow capacity of the river and its floodplain, at

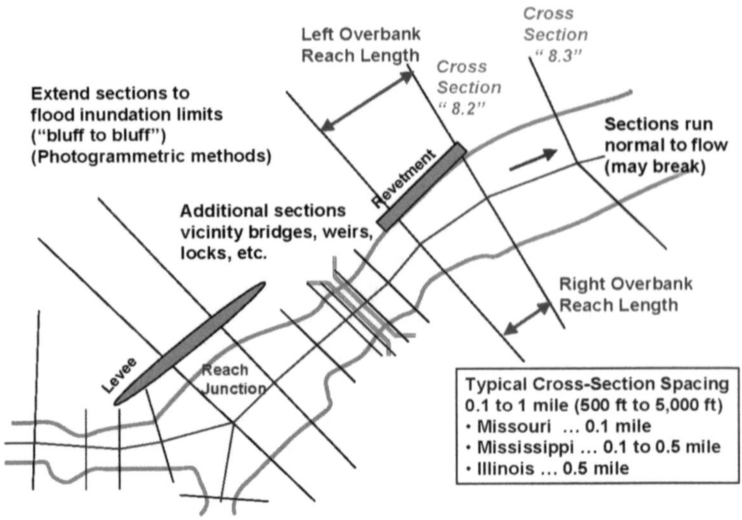

Fig. 3.5 Typical cross section configurations (*Source* EM 1120-2-1003 2002)

locations where there are changes in discharge, slope, shape, roughness, and at locations where the dykes begin and end (Fig. 3.5).

Cross sections should be placed at representative locations to describe the changes in geometry. Additional cross sections should be added at locations where changes occur in discharge, slope, velocity, and roughness. Cross sections must be also added at levees, bridges, and other structures. For example, if extensive flood inundation studies are analyzed, then the cross sections should be extended longer into the floodplain. Cross section elevations which have been constructed using any model based on a DEM are not so accurate compared to the cross sections which are measured in the field or digitized directly from photogrammetric stereo models (EM 1120-2-1003 2002). Bed slope plays an important role in cross section spacing. Steeper slopes require more cross sections. When there are not enough cross sections, they are spaced far apart and if the changes in hydraulic properties are great, the solution can become unstable. In general, cross sections spaced too far apart will cause additional numerical diffusion, due to the derivatives with respect to distance being averaged over too long a distance. Also, if the distance between cross sections is too far, such as when the courant number is much smaller than 1.0, then the model also becomes unstable. As cross section spacing is increased, the hydrograph will show some numerical attenuation/diffusion.

If the cross sections are too close together, then the derivatives with respect to distance may be overestimated (computed as steeper slopes than they should be), especially on the rising side of the flood wave. This can cause the leading edge of the flood wave to over-steepen to the point at which the model becomes unstable. To create a cross section, the cut lines will first be applied. These cut lines should always be considered perpendicular to the direction of the flow and start from the left to the right bank.

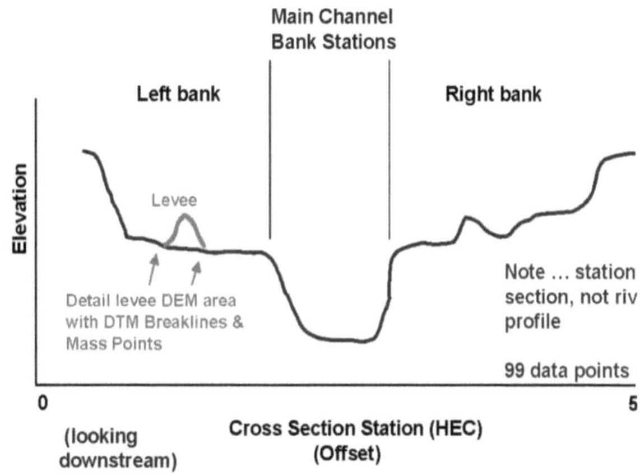

Fig. 3.6 Cross section convention for typical river and floodplain coverage (*Source* EM 1120-2-1003 2002)

The station-elevation data were extracted along the cut line from the DEM. ArcGIS 9.3.1 was used as support software for spatial processing of input and output data. The location and extent of cross sections were represented by the cross sectional cut lines layer in HEC-GeoRAS. The interval between cross sections was 150 m, with a total river section of 10 km in length. The cross sections which were created in HEC-GeoRAS were directly imported to HEC-RAS. In HEC-RAS, there is the possibility to interpolate the cross sections every 20 m. In order to create a more reliable flood modeling result (2D) in HEC-RAS, the extension of the cross sections in the floodplain area was considered mandatory. Ten cross sections were applied along the river and the data were manually imported to SOBEK. The schematic shape of a cross section, as a Y–Z profile created with HEC-GeoRAS (Fig. 3.6), was applied in HEC-RAS and SOBEK.

3.6 Discharge Data Modeling

Hydrometric data including hourly discharge (m^3/s) and water height (m) were defined for both models. The oldest severe flood occurred on 14 May 1957 with a maximum discharge of 480 (m^3/s), and the nearest flood was between 29 and 30 May 2008, with a maximum discharge of 205 (m^3/s). Based on available hourly discharge, the rating curve was applied between hourly height of water and discharge to calculate the missing discharge value for each month.

3.7 Roughness Values

The surface roughness values are some of the most important aspects in any flood modeling. Since it is difficult to measure surface roughness, the effectiveness of the specified roughness values was evaluated. Different combinations of roughness

Table 3.4 Land cover classes and surface roughness coefficients [49]

Land cover classes	Surface roughness coefficients
Coniferous forests	0.1
Broad-leaved forests	0.1
Natural grassland	0.35
Arable land/permanent crops	0.35
Pastures	0.035
Bare rocks	0.029
Black marls	0.029
Urban fabric	0.012
Airdrome	0.012
Water course	0.02–0.07
Marshes and water bodies	0.035

values were simulated to achieve the satisfactory simulation results [48]. The Manning equation is presented as follows:

$$V = \frac{R^{2/3}\sqrt{S}}{n} \tag{3.16}$$

$$\text{Hydraulic radius}\, R = A/P = w_{hi}(w + 2h) \tag{3.17}$$

In this research, based on physical characteristics of channel and floodplain areas, the Manning's coefficient based on Chow [49] was considered as a reference. Depending on the land use map, the values are shown in Table 3.4.

The table represents the surface roughness values used for flood simulation. The roughness coefficient in the channel (river bed) varied between 0.02 and 0.07 to calibrate models based on the optimization of roughness values in the channel and beds of the river.

3.8 Calibration Procedures

Generally, hydraulic models are calibrated by reducing the difference between calculated and measured flow conditions. According to Visser [50], the differences between calculated and measured flow conditions are reduced by expert judgment with regard to the effect of various specified hydraulic roughness components on the flow conditions, or by a recalculation of the flow conditions with adapted hydraulic roughness values. Optimization of the model needs a clear agreement or disagreement between the observed and modeled data [11]. Muthukishnan [51] also defined calibration tests as a comparison between the values of simulation with observed outputs.

One of the systematic methods was developed by Wijbenga et al. [52]. He developed the basis of a function ministration using the simple method as proposed by Nelder and Mead [53]. The calibration method is illustrated in Fig. 3.7.

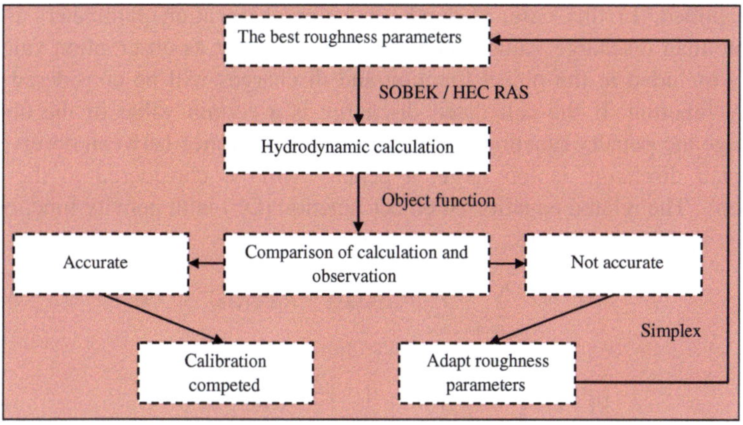

Fig. 3.7 Schematic calibration procedure Wisser et al. [54]

The algorithm is very simple. The function to be minimized represents the differences between calculated and measured water levels. The object function is based on a function of the roughness parameters which are calibrated. When the object function is minimal, it means that the roughness parameters are in an optimal case.

To prevent the roughness parameters from taking on unrealistic and inaccurate values, lower and upper bounds for the parameters are defined ([52–55]).

The calculation is based on a file with roughness parameters. Then, the calculated water depth and discharges are compared with measurement parameters using an object function. The simplex algorithm is then used to minimize the object function. In this step, the new roughness parameters are applied and the procedures of calculation start again. The process continues until the calculated measurements reach an acceptable agreement with measurement parameters and the desired accuracy is obtained. The selected object function is very important and the following parameters should be considered:

- First, it has to be decided which measurements are to be taken into account (stations, time series, etc.).
- Second, it has to be decided if all the measurements are equally important or if weight functions will be used.
- Third, it has to be decided in which way differences between measurements are quantified (e.g., absolute values, quadratic differences).

Wisser et al. [54] have suggested a penalty function as another option which has to be included. In this case, they mentioned that the object function consists of the sum of the differences between calculations and observation parameters and another penalty part. If a certain condition is fulfilled, this part is zero. On the other hand, if that condition is not fulfilled this part is positive [54]. Wisser et al. [54] described the penalty function in a simple way: "If both depth of water and discharges are included in the calibration procedures, then a penalty function is often applied. In this case, if depth of water observation parameters is more accurate than discharge values, then the depth of water as observation values are usually included in the object function and discharges will be considered as the penalty function. If the calculated discharge is a certain value of the observed discharge the penalty function is zero, but if the difference between observed and calculated discharge is too large, a penalty will be considered to the object function". The related equation of object function (OF) with penalty function (PF) is given by [54] as follows:

$$\mathrm{OF} = \sum_{\text{Stations}} \sum_{\text{time}} \left| h_{\text{calculated}} - h_{\text{observed}} \right| + \mathrm{PF} \tag{3.18}$$

$$\mathrm{PF} = \begin{cases} 0 & \text{if } \left| \frac{Q_{\text{calculated}} - Q_{\text{QH}}}{Q_{\text{QH}}} \right| < 0.05 \\ 10{,}000 & \text{else} \end{cases} \tag{3.19}$$

where:

h is depth of water, Q is discharge, and Q_{QH} is the discharge obtained from a discharge-stage.

In this research, calibration analysis was limited to the observed depth of water found in the literature and field reports on the 2008 flood. The adjustment was performed using different friction values in the channel. Manning friction coefficients were assigned based on different land use types to generate a friction surface [11]. The calibration of the model was based on the surface roughness parameters mainly within the channel. But for floodplain and overland flow modules the value was generated to produce real-world situations based on Chow [49].

3.9 Hydrodynamic Modeling

3.9.1 Basic Theory of the Models SOBEK and HEC-RAS

In this study, the SOBEK (1D–2D) and HEC-RAS (1D) modeling approaches were selected as hydrodynamic modeling because these two models were freely available for this research. In the HEC-RAS (1D) and SOBEK (1D–2D) models, the hydraulic equations of mass conservation and full momentum are solved so as to simulate the flood. In HEC-RAS (1D) the total discharge is distributed to the channel and the floodplain. In SOBEK (1D–2D), it is assumed that networks are two independent map layers, with the 2D network map layer overlaying the 1D network. The computational code determines the connection between the 1D and 2D layers, based on the map coordinates for the center of the 2D grid cell and the 1D connection node/calculation node. The flood simulation and model results provide the authorities with relevant information on how the flood will behave at the location where people live and how the flood will affect them [56].

3.9.2 Modeling Approaches

There are different types of flood modeling approaches:

1D model: In the 1D model, the principle is that the water level, velocity, and discharge only change parallel to the stream direction, and flow characteristic variation in any other direction is ignored. These models strongly depend on the geometric data at selected locations and floodplain and assume that river geometry for selected river sections only changes linearly along these sections.

2D model: In the 2D model approach, the principle is that the flow propagations change along two flow directions (x and y directions). The main advantages for using 2D model approach is that the flow in heterogeneous areas such as floodplains may be better represented in simulations, but flows across wide river beds are also not only in one direction but they need a more detailed representation of cross sections. Consequently, there are some problems with surface roughness parameterization which causes the calibration of 2D models to be relatively complex [57]. The 2D floodplain or overland flow module is based on a solution for the continuity and momentum equations.

1D–2D model: A coupled 1D–2D model approach is when the channel flow is simulated by a 1D representation and floodplain flow is simulated by a 2D representation, which is expected to benefit from advantages of both 1D and 2D models. The 1D–2D model is integrated by a 1D channel flow module with a 2D overland flow module. In the 1D–2D model, the 2D floodplain layer is overlaid with a 1D river layer, where layers are geometrically connected through map coordinates at the center point of a 2D floodplain grid element and the center points of 1D river section. The center points of the floodplain grid layer and the center points of the river grid layer lead the calculation algorithm in the numeric calculation procedures. In this case, the connectivity between channel and floodplain allows for overtopping of the river to floodplain inundations in case of a big flood occurrence [57].

3.9.3 HEC-RAS: Calculation Procedures

The 1D modeling approach is based on the 1D solution of the De Saint–Venant equation [58], such as the HEC-RAS model developed by the US Army Corps of Engineers. When considering channel flow, it is assumed that the flow behavior can be satisfactorily described as unsteady flow (flow characteristics may change over time) in one spatial dimension by two state variables: velocity (u) and water depth (h) as functions of time (t) and space (x) [59].

To solve u and h, two independent equations are required and usually the continuity equation (based on the conservation of mass principle) and the momentum equation (based on the conservation of momentum principle) are applied. The equations derived by De Saint–Venant in 1871 are shown below in their Eulerian form per unit width of channel with no lateral inflow [3]:

$$\frac{\partial(h + z_b)}{\partial t} + \frac{\partial(uh)}{\partial x} = 0 \tag{3.20}$$

$$\frac{\partial u}{\partial t} + u\frac{\partial u}{\partial x} + g\frac{\partial(h + z_b)}{\partial x} + c_f\frac{u|u|}{h} = 0 \tag{3.21}$$

where:

x is the position along the channel axis (m); t is time (s), u is the velocity vector in the x-direction (m/s), h is water depth (m), z_b is the local bottom level above the reference date, c_f is the dimensionless bottom friction coefficient, and g is the constant of gravity (9.81 m^2/s).

The first equation represents the rate of volume stored over unit length of a channel, and the discharge which changes along the channel per unit of time. The second equation expresses the change of momentum in a controlled volume of unit length of a channel and reflects the inertia of the water mass present in that controlled volume [55].

In HEC-RAS, the total discharge is distributed to the channel and the floodplain elements as per their conveyance. The governing equations for both the floodplain and the channel grid elements are as follows:

$$\Delta Q_T + \frac{\Delta A_c}{\Delta t}\Delta x_c + \frac{\Delta A_f}{\Delta t}\Delta x_f + \frac{\Delta S}{\Delta t}\Delta x_f - \overline{ql} = 0 \tag{3.22}$$

$$\frac{\Delta(\mathrm{Q_c}\Delta \mathrm{x_c} + \mathrm{Q_f}\Delta \mathrm{x_f}}{\Delta \mathrm{t}\Delta \mathrm{x_e}}\Delta(\beta V_T Q_T) + g\bar{A}\left(\frac{\Delta Z}{\Delta x_e} + \overline{S_f}\right) = 0 \tag{3.23}$$

where

Q is discharge (L^3T^{-1}); A is cross sectional area (L^2) = $\Phi K_c/(K_c + K_f)$; K is conveyance; V is velocity (L^{T-1}); Z is water level (L); S_f is friction slope; S is storage term (L^2); $\overline{S_l}$ is average lateral inflow (L^3T^{-1}); β is coefficient and Δx_e is equation flow path (L). The parameters of C, f, and T are terms for channel, floodplain, and combined channel and floodplain, respectively. In the second equation, the friction slopes for both the channel and the floodplain are assumed to equal zero.

As mentioned before to create all necessary input data for HEC-RAS modeling, the extension of Hec-GeoRAS model in ArcGIS was applied. The process modeling in HEC-GeoRAS and HEC-RAS modeling is presented in Fig. 3.8.

Froude Number (Fr): In the HEC-RAS model (1D), the Froude number is an important parameter which deals with the characteristics of the open channel flow. The Froude number is defined as the ratio of flow velocity to the velocity of free surface wave at a specific location. The Froude number in flood simulation models is an important factor in both: (1) Understanding the flood wave movement and (2) The numerical stability of the model. It could also be considered as a ratio of inertial forces to the gravity forces, and is expressed in the following equation [60–63]:

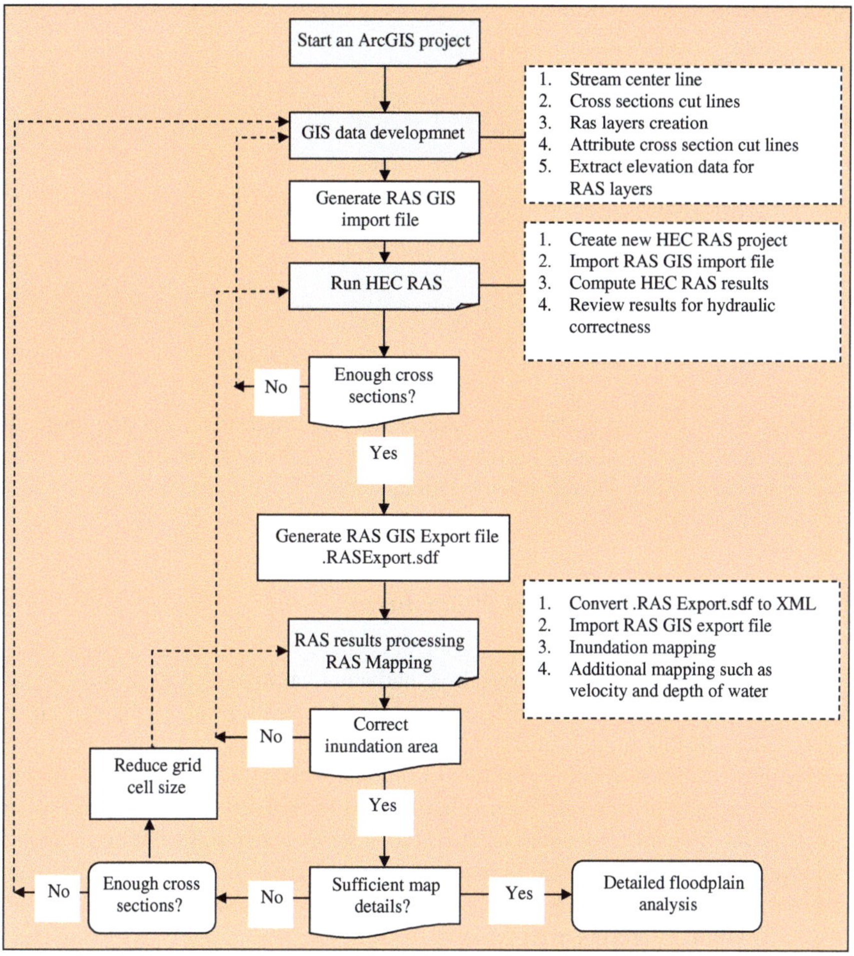

Fig. 3.8 The process modeling in HEC-GeoRAS and HEC-RAS (HEC-RAS manual [64])

$$\mathrm{FR} = \frac{V_u}{\sqrt{gL_c}} \tag{3.24}$$

where,

V_u = upstream velocity and L_c = characteristics length for the channel flow.

The dimensionless Froude number (Fr) is applied to characterize the flow regime, where:

- Froude Number (Fr) < 1 represents subcritical flow
- Froude Number (Fr) > 1 represents supercritical flow
- Froude Number (Fr) $= 1$ represents critical flow (Fig. 3.9)

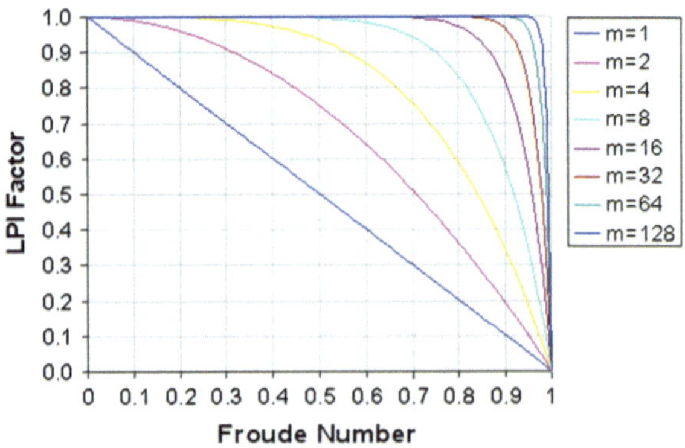

Fig. 3.9 Froude number (Fr) = 1 represents critical flow (HEC-RAS manual)

Ubaye River is composed of both natural and man-made channels (the upstream is channelized, the middle and downstream parts are natural), therefore, a critical flow scenario was considered for modeling.

3.9.4 SOBEK: Calculation Procedures

Within the 1D–2D modeling approach, a mathematical model consists of a 1D–2D flow model [65] such as the SOBEK model developed by Delft Hydraulics in the Netherlands. In this model, unsteady flow discharge varies as a function of time and therefore, all the hydraulic factors of a cross section change, such as velocity varying and depth of water as a function of time [3]. SOBEK is a model that solves the De Saint–Venant flow equation in 1D as long as water stays within its banks. However, when dyke/levees are breached or overtopped the model automatically switches to 2D [66]; SOBEK User's manual [67], www.sobek.nl).

The equation is presented as follows:

$$\frac{\partial A_t}{\partial t} + \frac{\partial Q}{\partial x} = q_{\text{lat}} \tag{3.25}$$

$$\frac{\partial Q}{\partial t} + \frac{\partial}{\partial x}\left[\frac{Q^2}{A_f}\right] + gA_f\frac{\partial A}{\partial x} + \frac{gQ|Q|}{C^2RA_f} - W_f\frac{\tau_{wi}}{p_w} = 0 \tag{3.26}$$

where:

A_f is the wetted area (m^2); q_{lat} is the lateral discharge per unit length (m^2/s); Q is the discharge (m^2/s); t is time (s); x is the distance (m); g is the gravity acceleration (m^2/s) = 9.81; h is the water level (m); C is the Chezy coefficient (m$^{1/2}$/s); R is the hydraulic radius (m); W_f is the flow width (m); T_{wi} is the wind shear stress (n/m^2); P_w is the water density (kg/m^2) = normally 1,000.

Forces caused by bed friction and earth gravity usually primarily determine flow conditions and other forces are less important. During computation, the

Chezy coefficient C may be determined in a number of ways, referring to various formulations, e.g., Chezy, manning. In particular, if a Manning value is provided, the Chezy coefficient is calculated as:

$$C = \frac{R^{0.125}}{n} \tag{3.27}$$

SOBEK overland flow consists of a 2D modeling system based on a subset of the shallow water equations as follows:

$$\frac{\partial h}{\partial t} + \frac{\partial (uh)}{\partial x} + \frac{\partial (vh)}{\partial y} = 0 \tag{3.28}$$

$$\frac{\partial u}{\partial t} + u\frac{\partial u}{\partial x} + v\frac{\partial u}{\partial y} + g\frac{\partial (h + z_b)}{\partial x} + C_f\frac{u\sqrt{u^2 + v^2}}{h} = 0 \tag{3.29}$$

$$\frac{\partial v}{\partial t} + u\frac{\partial v}{\partial x} + v\frac{\partial v}{\partial y} + g\frac{\partial (h + z_b)}{\partial y} + C_f\frac{u\sqrt{u^2 + v^2}}{h} = 0 \tag{3.30}$$

$$V = \left(u^2 + v^2\right)^{1/2} \tag{3.31}$$

where, x and y represent the orthogonal axis; u and v the velocity vectors along these axis, respectively. The staggered grid shown in Fig. 3.10 shows that water level (h) and flow velocity vectors (u and v) are computed at alternating grid points. The flow velocity (v) can be computed as the vector sum of the vector velocities u and v.

The SOBEK software package integrates a 1D modeling package with a 2D hydrodynamic prediction package. Thus, it has the functionality of including the one-dimensional river/channel flow and two-dimensional overland flow, which is called "SOBEK 1D–2D" ([66]; SOBEK User's Manual [67], www.sobek.nl). Figure 3.11 represents the modeling process in SOBEK.

The SOBEK software package integrates a 1D modeling package with a 2D hydrodynamic prediction package. Thus, it has the functionality of including the 1D river/channel flow and 2D overland flow, which is called "SOBEK 1D–2D" ([66]; SOBEK User's Manual [67], www.sobek.nl). Figure 3.6 represents the modeling process in SOBEK.

The data input and editing interface in SOBEK is called NETTER. NETTER offers the possibility to set up the schematization and also offers advanced analysis tools to show model results. NETTER has two editing models: (a) for setting up the schematization (e.g., by editing new nodes); and (b) for editing the attribute data. NETTER also provides two separate menu bars to input and edit 1D feature (drainage channel) and 2D grid layers representing floodplains ([66]; SOBEK User's Manual [67]).

Before starting schematization it is important to have DEM imported to the NETTER, which was developed in ArcGIS, and imported directly to the schematization environment as an Ascii format using "2D grid" tools. Then, the flow

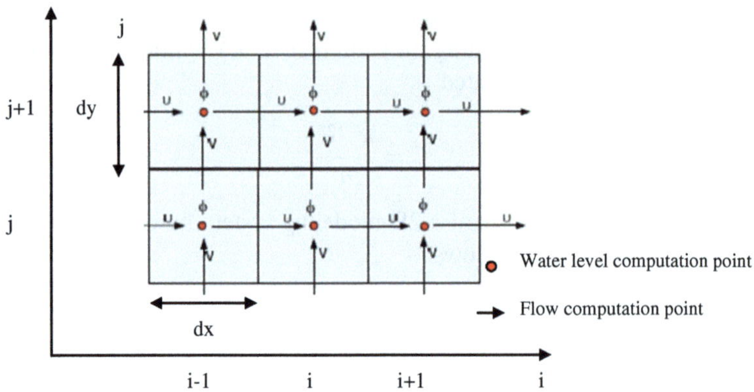

Fig. 3.10 Staggered grid schematization for 2D flow simulation [142]

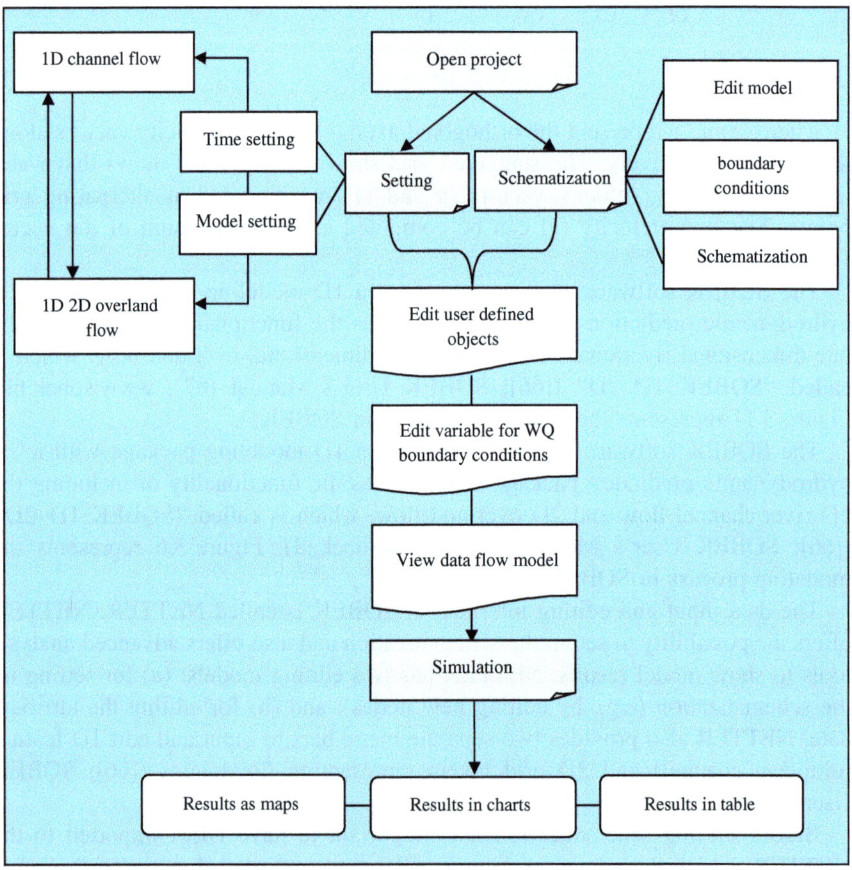

Fig. 3.11 The process modeling in SOBEK (Based on SOBEK manual)

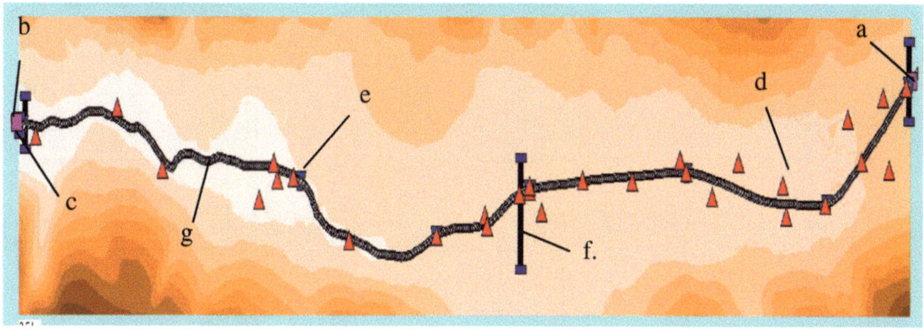

Fig. 3.12 Schematization of 2D network with history stations; **a** upstream boundary condition (1D), **b** downstream boundary condition (1D), **c** downstream boundary condition (2D), **d** history stations, **e** cross sections, **f** line measurement, **g** nodes

boundary conditions should be defined in 2D schematization. The flow boundary conditions will be connected using a flow connection node in order to define the 1D component. Figure 3.12 shows the schematization environment in SOBEK. To determine the exact value of parameters, such as depth of water in time on specific locations in 2D grids, the history stations nodes were defined.

3.9.5 Data Requirements, Processing, and Preparation: Setting Up the Models

For applying the distributed hydrologic model for any purpose, it is necessary to prepare a huge amount of spatial and temporal input data. The importance of various input data varies according to the application of the specific purpose. For example, in flood inundation simulations, topography layer is one of the most important spatial input data which deals with the flow behaviour in overland flow and floodplain areas. In any application of a hydrodynamic model, the drainage network is delineated from the topography data or a DEM.

As the drainage network is the most critical item in surface flood simulation, it is important that it is delineated with more reliability and accuracy. Ubaye River was constructed in SOBEK (1D–2D) and HEC-RAS (1D) using the following information/data:

DEM: It covers channel and floodplain areas of approximately 14 by 10 km, with a 5 m resolution. The DEM is the digital representation of natural topographic as well as man-made features located on the surface of the earth [1] which affects the flow of water, such as dykes, embankments, and buildings. Many of these features have to be extracted from topographic maps, building footprint maps, field surveys, bathymetry data, and other sources [3]. Due to lack of high resolution of DEM across the Barcelonnette area, man-made features are added to the DEM to create the final digital elevation surface map to use in hydrodynamic simulation.

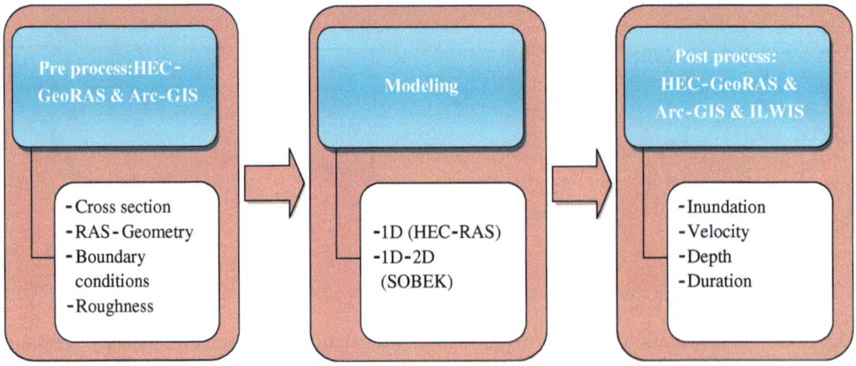

Fig. 3.13 Input and output information in HEC-RAS and SOBEK modeling

Roughness value: An estimation of channel and floodplain roughness coefficients.

Geometric data: Modeling input data are included in spatial data and temporal data. The primary geometrical inputs required for 1D modeling were flow path, center line, main channels, banks, and cross section cut lines. The optional geometrical input was the application of left and right dykes in urban areas. In order to model flood propagation, a series of input data created with the HEC-GeoRAS model were applied in HEC-RAS, such as cross sections of a Y–Z profile. The location and extent of these cross sections were represented by the cross sectional cut lines layer in the HEC-GeoRAS model, an ArcGIS extension. Figure 3.13 shows a schematic representation of the necessary input data required for both modeling and output information as a result of the simulation.

Boundary conditions: They were defined in the form of upstream hydrograph data and downstream water level. Boundary conditions were applied to define the inflows and outflows at the model boundary.

Alemseged and Rientjes [57] defined the boundary conditions as in commonly specified terms as inflow and outflow elements of the model domain. In both modeling approaches, the same discharge hydrograph was used for the upstream boundary condition. Due to lack of data and no available recorded downstream data, an imaginary lake condition was assumed in the SOBEK model, which is set to have a water holding capacity of up to 2 m from its immediate value of elevation. To preserve the lake conditions, a constant water level was specified and a free flow was allowed throughout the simulation [11].

It is assumed that it will constrain the flow in the model and could eventually even create a backwater effect, as it seems the water returned back again to the watershed. The ideal downstream boundary condition would be a rating curve that reproduces the relational discharge and water level for any situation. In particular, with detailed data, the HEC-RAS allows to include a rating curve, stage, and flow hydrograph and the slope of river course as a downstream boundary condition. Figure 3.13 represents the schematic figure of modeling processes and some input and output information in HEC-RAS and SOBEK modeling.

3.9.6 Output Data

A lot of information for hazard assessment is contained within the time series of maps with water depth, flow velocity, duration, shear stress, etc. To analyze these parameters an aggregation procedure has been developed to create the parameter maps that describe the different aspects of the flood event in the Barcelonnette area. These parameters are explained as follows [68]:

Hydraulic depth (unit: m): In order to create depth mapping, assumptions must be made on the nature of flows on floodplains. The model results can be used to map water elevations onto a DEMof the catchments. This map shows the maximum depth that occurred during the inundation. The rationale behind this parameter map is that areas with deep water are more dangerous to people and potentially more damaging to objects like houses and cars. It identifies areas where the second floor, or even the third and fourth floor of houses, are not safe refuges. The maximum water depth map also serves as a possible means for model calibration. Maximum water depth is one of the few flood parameters that can easily be retrieved after a flood event because of the wetting marks in and on structures.

Flow velocity (unit: m/s): This map shows the maximum flow velocity that occurred during the inundation. The rationale behind this parameter is that velocity is a component of the floodwater that can sweep people off their feet and make cars float away. This map shows where preferential flow paths may develop that could be dangerous for children, adults, and cars.

Duration (unit: h): This map estimates the time the floodwater remains at a certain location. It is based on several assumptions regarding the drainage of the floodwater from the flooded area. The rationale behind this parameter is that it gives a first, rough impression of how long the floodwater will stay in the area.

This is the minimum time period that people have to be relocated, that businesses and industries are closed, and that transportation in and throughout the area might be impossible or hindered. It is a strong parameter with which to assess the economic and social impact of the flood on the people living and working in the area.

It is also an important parameter for estimating agricultural damage because many crops, like fruit bearing trees and vineyards, can withstand inundation of their stems for a short time (usually some days), but if the period becomes too long, the roots will starve from oxygen depletion and the trees will die.

Shear stress (t): In the case of open channel flow, it is the force of moving water against the bed of the channel. This force induces "bed shear stress" which can be determined from the following equation [69]:

$$\tau_0 = \gamma R S_0 \tag{3.32}$$

where,

τ_0 = bed shear stress; γ = specific weight of water; R = hydraulic radius; and S = bed slope.

In terms of bed velocity, shear stress can be expressed as:

$$\tau_0 = \gamma(Vn)^{3/2} S^{1/4} \tag{3.33}$$

where,
$V =$ bed velocity; $n =$ Manning's coefficient.

3.10 Scenarios

In order to have valuable and basic information available for any future management, it is necessary to provide the authorities, stockholders, other scientists, and the public with flood hazard maps which are basic but which contain fundamental information for any environmental, technical, economical, and political decisions regarding flood event management. This shows the potential adverse consequences associated with different flood scenarios for civil protection purposes, including information on potential sources of environmental damages as a consequence of floods [70].

3.10.1 DEM Scenario

Topography is considered an important land surface characteristic that affects the flow distribution over the channel and floodplain in the catchment. According to Vaze [1] and other investigators, such as Abbott et al. [71] and Rohdenburg et al. [72] all of spatially distributed hydraulic and hydrological models use topography as a DEM to derive bathymetry (Channel). Alkema [73] also mentioned that in hydraulic modeling, the output is affected largely by model input parameters, such as the DEMs and other topography related properties such as slope gradients, slope aspects, and drainage density.

In flood modeling, it is a great interest to know how much the resolution of DEM determines and affects on the model outputs. Researchers have also realized that DEM quality and resolution significantly affect the accuracy of any extracted hydrological features and drainage networks delineation [74].

The indirect effects of DEM on hydrodynamic simulation's results can be seen on the model inputs, such as the hydraulic roughness, which are simulated at the scale of the DEM elements. Hydraulic roughness can change with DEM resolution and, as such, also affect simulation results.

Almost all of the applications including water resource management and hydrological modeling required a high quality of DEMs because the accuracy of DEMs does affect the accuracy of hydrological predictions. There are numerous studies [47, 75–77] reported in the literature which compared spatial indices derived from different resolution DEMs (e.g., 100 and 10 m grid cell resolution).

Table 3.5 Advantages and disadvantages of low and high resolution of DEM

DEM	Advantages	Disadvantages	Sources
Low	Takes a short time for simulation; in some flat areas or on a global scale, it is adequate to capture	Missing a larger number of feature surface information such as dyke; delineation of drainage networks in lowland flat terrain	[47, 75, 76, 140, 141]
High	Small error; benefits the local scale, contains most of man-made feature surface	Takes a long time for simulation	[47, 75, 76, 140], [73]

Table 3.5 represents some advantages and disadvantages of low & high resolution of DEM according to the experience of some investigators.

According to Alkema [3], the computation time of simulation depends on the number of cells in the grid and consequently, a large number of grid cells for the large area make it impossible to feed it directly into the flood model, because the calculation time would become enormous.

Vaze [1] mentioned that the vertical accuracy of grid cell elevation is a critical and important factor, as a small error in grid cell elevation can result in totally different and incorrect model predictions of the values of the spatial indices derived from DEM. He also indicated that the landscape process scale (global, regional, local) is the key driver in determining a useful and more applicable grid cell resolution scale.

The landscape process scale and the resolution of DEM have been discussed in many papers. For example, Gallant and Hutchinson [79] point out that the grid resolution of DEMs can profoundly influence the spatial patterns of attributes and also the related models built from these attributes. Schoorl et al. [80] also discussed the influence of implications of varying DEM resolutions on the numerical values of attributes derived from them.

In another example, Warren et al. [81] compared slopes measured in the field with different resolutions of DEMs, and found that higher DEMs with 1 m resolution produced much better and more accurate results than lower DEMs with 12 m resolution. They commented that this variation can lead to widely varying estimations in environmental parameters such as soil erosion. As a general message, Wilson and Gallant [82] stated that "additional work is required to identify the important spatial and temporal scales and factors that influence or control the process and patterns operating at particular scales".

Therefore, it is essential to quantify the impact of different resolution DEMs application on hydrological modeling as an important variable which affects the accuracy and reality of the results as the simulation is moved from a high resolution to a lower resolution in comparison with observed and measured data Vaze [1]. In this study, in the case of the DEM scenario, the aim was to compare different resolutions of DEMs (5 m against 10 m) to understand how the grid resolution of a DEM affects the flow characteristic results. To compare the flow propagation results, depth of water and velocity were estimated and simulated in the HEC-RAS model based on two different DEM resolutions. To this purpose, the hydrograph of flood events of 2008 was used because there was the chance to calibrate the model based on observed information from that event.

3.10.2 River Change Scenario

In the river change scenario, the river morphology was first digitized according to available aerial photos for different years. Six series of vertical aerial photographs were used: 1956, 1974, 1982, 1995, 2000 and 2004. These series of photographs were applied to historically and quantitatively assess the area occupied by different land cover in the whole study area, and to get a feeling about the river channel changes over the time. Then, in hydrodynamic modeling, different hydrographs were applied to approximate how river channel changes affect flow propagation.

3.10.3 Dyke Scenario

A high percentage of the world's population lives close to oceans, seas, lakes, and rivers. In many cases, the life of these people, coupled with their personal property and belongings, depend on flood defense systems [83]. In 2004, a commission of the European communities reported that river floods may occur whenever the capacity of the channel or man-made drainage system is unable to cope with the volume of water generated by rainfall or when flood defenses fail [70]. In recent years, extreme river floods across the world and all over Europe occurred due to the breaking of these flood defense systems and the protection of infrastructures. For example, during the August 2002 floods, more than 130 dyke breaches occurred in Germany along the Elbe river which caused a total damage of 15 billion € [84–87]. In another example, in the Netherlands, the floodplain area has often historically been faced with the threat of several floods from the sea, as well as from rivers, which were mostly due to flood defense failure Möllmann et al. [87]. Also, the failure of dykes around the city of New Orleans during Hurricane Katrina turned a bad hurricane into a major urban catastrophe. Major damage was due to the dyke failure along the Industrial Canal [88]. Usually, floods have high damage potential due to the growth of urban areas or industrial sites in floodplains. In many cases, these areas are protected by river dykes. Depending on the characteristics of the river, the floodplains, dykes, and characteristics of the dyke breach, such as the location and width of the breach, significant volumes of water may spill over into the dyke hinterland, reducing the peak of the flood wave downstream of the breach location [84, 89–91]. River beaches protected by dykes exhibit high damage potential due to a high value accumulation in the hinterland areas. While providing efficient protection against low magnitude flood events,

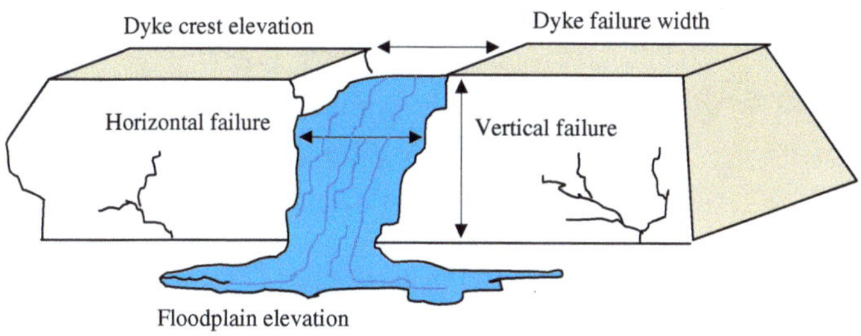

Fig. 3.14 Simplified dyke failure in a schematic representation

dykes may fail under the load of extreme water levels and long flood durations. Losses arising from subsequent inundation may be dramatic, not only because of the high value concentration in the dyke protected floodplain, but also because of the fast water level rise and high flow velocities caused by rapid breach outflow [2]. Dyke failure occurs in different conditions:

- Dykes can fail by overtopping or by being inundated at a prescribed elevation for a specified duration.
- Dykes can fail to a prescribed elevation different from the floodplain elevation.
- Specify dyke breach width and,
- Specify a rate of failure both vertically and horizontally (Fig. 3.14)

The dyke scenario simulation results are beneficial for the authorities, and planners should know about the possible effects of failure, finding critical places which are more vulnerable before an extreme flood occurs in the case study. In this study, in case of dyke failure, the aim was to find how strong the dykes and channel modification are in both the city and natural sectors of Ubaye River. In this context, simulations of different failure scenarios applying various hydrograph propagations are useful to investigate the river behavior during flood events. Ubaye River is divided to two parts; modified/channelized and natural river bed. These two parts are included in a channelized part of the city and the undisturbed river bed without any infrastructures. The dykes are located in urban reach of the river having an approximate height of ~2 m and a length of ~3.5–~4 km Ramesh [92]. Potential affected areas from damages in different scenarios could be estimated by comparing the results of the simulation derived from different dyke scenarios with available aerial photos.

Dyke scenarios were applied to the situation with and without dykes, giving a rough idea of the flood defense effects of dykes. HEC-RAS hydraulic modeling was used to execute unsteady flood river routing of dyke breach hydrographs through the upstream and downstream channels. SOBEK hydrological modeling was applied to estimate the flow propagation in floodplain in case of dyke breach and overtopping of the banks and inundation occurrence. Input data such as RAS-geometry was achieved with HEC-GeoRAS (ArcGIS extension) and additionally, ArcGIS 9.3.1 was applied for spatial processing of the input and model of data, and to visualize the outputs. The applied analyses include the following scenarios:

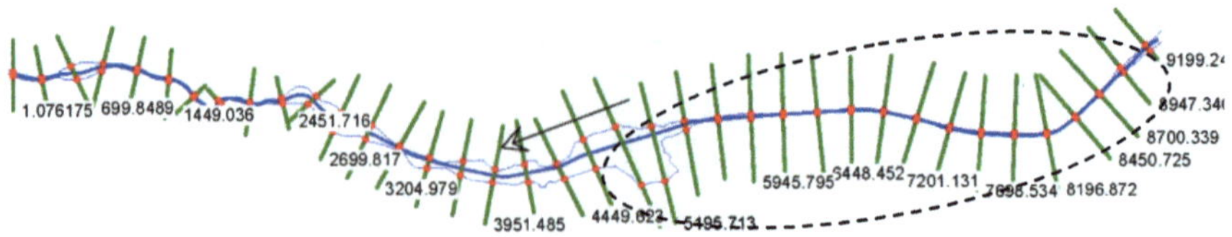

Fig. 3.15 Dyke position parallel to the river in the city part of the river

scenarios (1) and (2): modeling of the flood event in 2008 (with maximum discharge of 205 m^3/s) with and without the dykes, respectively, supporting scenario (3): simulation of the flood event in 1957 (with maximum discharge of 480 m^3/s) considering the dykes and the scenario (4): modeling of the flood event in 1957 (with maximum discharge of 480 m^3/s) for the river profile of 2008 considering the dyke. The Dyke breach case is also performed as the worst situation in the last scenario. The dyke position was introduced in the DEM using HEC-GeoRAS and an elevation of 1.5 m (before reconstruction) and 2 m (after reconstruction) were defined in HEC-RAS for the years of 1957 and 2008, respectively (Fig. 3.15).

The left and right banks of the Ubaye River in the city area were rebuilt after an extreme flood in 1957, according to state of the art engineering.

3.10.4 Model Scenario

Different model simulations in the context of modeling results have also been discussed in many papers [48, 93]. Horritt and Bates [48] evaluated the flood simulation results as obtained from a 1D raster based model and a 2D model with finite element discretization.

In the model scenario, SOBEK as a 1D–2D model and HEC-RAS as a 1D model (Fig. 3.16) were applied in a hydrodynamic simulation to find any advantages and disadvantages of both modeling by comparing the outputs.

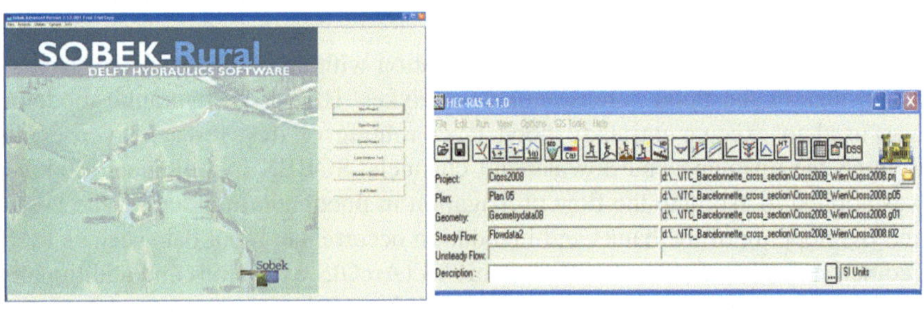

Fig. 3.16 Schematic shape of both model

3.11 Flood Frequency Analysis

Analyzing the extreme event is mostly impacted in hydrologic systems. For any design, river engineering and hydraulic planning is not enough to say that the maximum observed flood was, for example, 480 m^3/s, it is also necessary to consider the frequency of occurrence of the mentioned flood. Parodi [94] realized that the magnitude of an extreme event is inversely related to its frequency of occurrence e.g., very severe events have less frequency compared to more moderate events. Therefore, according to Parodi [94], "the frequency analysis is related to the magnitude of extreme events to their frequency of occurrence through the use of probability distributions". The hydrologic data in frequency analysis are assumed to be independent and identically distributed. For this purpose, the annual maximum of the variable, e.g., the annual maximum discharge, should be considered.

These variables are the largest instantaneous peak flow occurring at any time during the year with the expectation that successive observations of this variable from year to year will be independent [84, 95–100]. According to Parodi [102], the results of flood frequency analysis can be used for many engineering purposes: for example, the design of bridges, dams, dykes and flood control structures; economical purpose of flood control projects; and for delineation of floodplains and to determine the effect of encroachments on the floodplain [94].

3.11.1 Return Period

Flood hazard assessments are usually based on a number of flood scenarios. Each scenario is associated with a probability P_E or return period T. In many countries around the world, such as Canada, UK, USA, Italy, France, Spain, Germany, and New Zealand, the floodplain area is affected by a 100 years return period flood that plays an essential role for flood mitigation strategies. In Europe, according to the proposed directive of the European Union on the assessment and management of floods, two flood scenarios with return periods of 10 and 100 years and an extreme scenario with a higher return period were defined [84].

Stedinger et al. [101] suggested that in flood hazard assessments two steps should be considered: the first is to estimate the T-year discharge along the river or water course, and the second is to transfer the discharge values into floodplain. The most common approach to estimate the T-year discharge along rivers is flood frequency analysis. Hosking and Wallis [102] have a discussion related to the scale. They concluded that in many cases, local frequency analysis is complemented by regional flood frequency analysis by applying the data from those gauging stations that have similar flood behavior or characteristics.

Apel [84] had referred to some limitations faced by the frequency analysis. For example, he mentioned that in any flood frequency analysis, the results are affected by insufficient data sets or missing data. Therefore, in many cases the discharge data series is no longer than 30–50 years. Consequently, estimation of floods with return periods above 100 years is an uncertain level.

He also mentioned that the 1D or 1D–2D hydrodynamic models approaches associated with certain discharge values can include the effects of dyke breaches. However, since the T-year discharge for certain river sections is taken from flood frequency analysis, the effects of upstream dyke breaches do not propagate and do not affect the flood frequency analysis at downstream gages [84].

Except for the return period calculation, another aim of flood frequency analysis in this research was to investigate how flood frequency distributions were influenced by dyke breaches. Therefore, with the assumption of dyke breaches, the flow propagation and flood behavior was also estimated. In return period analysis, it is supposed that an extreme event is defined to have occurred if a random variable X is greater than or equal to some level x_T. The recurring interval t is the time between occurrences of $X \geq x_T$ [94]. For each observation, there are two possible outcomes: either "success" $X \geq x_T$ (probability p) or "failure" $X < x_T$ (probability $1-p$). Since the observations are independent, the probability of a recurrence interval of duration T is the product of the probabilities of $t-1$ failures followed by one success, that is, $(1-p)\ t-1{\cdot}p$. Assuming that the series of data is infinite, the $E(T)$ can be expressed as:

$$E(t) = \sum_{t=1}^{\infty}(1-p)^{t-1}.p \tag{3.34}$$

Developing this expression in terms and after some algebra is shown in the following equation:

$$E(t) = T = \frac{1}{p} \tag{3.35}$$

So the probability of occurrence of an event in any observation is the inverse of its return period.

$$P(X \geq xT) = \frac{1}{T} \tag{3.36}$$

3.11.2 Gumbel Extreme Value Distribution

The objective of these distributions is to build the relation between the probability of the occurrence of a certain event, period, and its magnitude. For this purpose, first, the maximum discharge values were ranked values from low to high. So the first value was assigned to the lowest data value and the last one was assigned to the highest data value.

Then, for each observation the left sided probabilities were calculated by:

$$P_L = \frac{R}{N + 1} \tag{3.37}$$

where:

P_L is left sided probability (probability of having less value in the series); R is the rank; and N is number of observations.

The return period for each observation was determined using the following equation:

$$T = \frac{1}{P_R} = \frac{1}{1 - P_L} \tag{3.38}$$

The plotting position for each observation was determined as follows:

$$y = -\ln(-\ln P_L) \tag{3.39}$$

In the total 31 years of data, the maximum discharge per year was extracted from the originally observed data, and all the above process have been done in excel and the "CumFreq" program. The statistical methods discussed are applied to extend the available data, and hence predict the likely frequency of occurrence of a flood event.

3.12 Trend Detection in Hydrometeorological Variables

Hydrologists have been interested in trend testing for hydrometeorological variables such as precipitation, temperature, and stream flow. The trend analysis of hydrological and meteorological series is important and even more relevant when considering the regional effects of global climate change. Trend analysis provides an idea about climatic variability in the watershed. Trend analysis is done with Kendall's method to check the significance of trends. The Mann–Kendall test as a nonparametric method is often used in trend analysis and suggested by many investigators as a very strong method to detect trends. For example, Hirsch [103] applied this test in a water quality trend test application. In another example, Helsel and Hirsch [104] provided a comprehensive review of statistical approaches used for trend analysis of water recourses time series using Mann–Kendall. This test was originally derived by Mann [105] and developed by Kendall [106] subsequently. This part of the research is an attempt to develop an integrated methodology to analyze statistical significant trends in the time series of precipitation, discharge, and temperature data to monitor the changes over time.

3.12.1 Detection of Randomness

One of the problems in detecting trends in hydrologic time series is the effect of serial correlation. If there is a positive serial correlation in a time series, the non-parametric test will suggest a significant trend in the time series [107–110]. To avoid this problem, the autocorrelation structure of the time series should be investigated. If the time series is random, the autocorrelation coefficients are not statistically different from zero. In other words, the autocorrelation coefficients do not cut the confidence interval (CI) at any desired level of significance, i.e., 95 %.

The CIs are given by

$$CIs = \frac{Z_{\propto} - \propto/2}{\sqrt{n}} \tag{3.40}$$

where:

Z is defined as the percent point function of the normal distribution; n is the sample size and $\propto$ is the significance level.

3.12.2 Detection of Homogeneity

Different methods and tests have been introduced to test the homogeneity of hydroclimatic variables [111, 112]. A climatic variable is said to be "homogeneous" when its variations are caused only by fluctuations in weather and climate [113]. In order to detect homogeneity in monthly precipitation, maximum discharge series, and monthly temperature, the "run test" is applied in this study because it is a common and valid method suggested by many investigators [110]. In this test, time series of length n and x_{med} or the median of the time series are considered. A code called "a" was considered for any value $x_j > x_{\text{med}}$ and a code called "b" was considered for any value $x_j < x_{\text{med}}$. Each uninterrupted series of "a" and "b" codes is called a "run". The distribution of the number of runs (N) approximates a normal distribution with the following average (E) and variance (Var):

$$E(R) = \frac{N+2}{2} Var(R) = \frac{N(N-2)}{4(N-1)} \tag{3.41}$$

The Z statistic is defined as:

$$Z = \frac{R - E(R)}{\sqrt{Var(R)}} \tag{3.42}$$

For significance level of $\alpha = 0.01$ and $\alpha = 0.05$, the null hypothesis of homogeneity is verified if $|Z| \leq 2.58$ and $|Z| \leq 1.96$, respectively.

3.12.3 Detection of Trend: Mann–Kendall Test

The Mann–Kendall test was applied by many researchers and investigators to estimate the trend in climatological and hydrological parameters from the past to present [104, 107, 114–117]. This method as a nonparametric test identifies trends in time series data [118]. The Mann–Kendall method is widely used in environmental and natural science as a simple, robust method which can cope with missing values and values below a detection limit [110, 119]. The first original proposals of this test derived by Mann [105] and Kendall [106], the covariance between the Mann–Kendall statistics, were suggested by Dietz and Killen [120]. Then, the test was developed in order to include seasonality by Hirsch and Slack [103], multiple monitoring sites were proposed by Lettenmaier [121] and covariates representing natural fluctuations were considered by Libiseller and Grimvall [119].

Khambhammettu [122] realized that one of the advantages of this test is that the data does not need to conform to any particular distribution. Moreover, data as nondetects can be included as a common value that is smaller than the smallest measured value in the data set. Also, the data values are evaluated as an ordered time series. Each data value is compared to all subsequent data values. As mentioned above, the nonparametric Mann–Kendall rank correlation test [123] was applied to detect any possible trend in precipitation, discharge, and temperature series, and to test whether or not such trends are statistically significant [124]. This test, usually known as Kendall's τ statistics, has been used in hydrology and climatology to test randomness against trends of hydrologic time series. As it is a rank-based procedure, it is robust to the influence of extremes and a good test for skewed data. Kendall [106] reported a normal approximation test that could be used for data sets with more than 10 values, provided there are not many tied values within the data set. The procedure is as follows:

For the nonparametric Mann–Kendall method, for any sample of n variables, $x_1, \ldots, \underline{x}_n$, the null hypothesis ($H_0$) indicates that the sample is independent and distributed randomly. The alternative hypothesis of a two-sided test (H_1) also indicates that x_i and x_j are not identical for all k & $j \leq$ n with $i \neq j$.

The Mann–Kendall test is based on test statistic S, with a zero mean and computed variance from Eqs. (3.45) and (3.46) [125] as:

$$S = \sum_{i=2}^{n} \sum_{j=1}^{i-1} \text{sign}\ (x_i - x_j) \tag{3.43}$$

$$\text{sgn}(x_j - x_{jk}) = \begin{cases} +1 & \text{if}\ (x_i - x_j)S > 0 \\ 0 & \text{if}\ (x_i - x_j)S = 0 \\ -1 & \text{if}\ (x_i - x_j)S < 0 \end{cases} \tag{3.44}$$

Where:

The x_j are the sequential data values; n is the length of the time series; and sign $(x_i - x_j)$ is -1 for $(x_i - x_j) < 0$; 0 for $(x_i - x_j) = 0$, and 1 for $(x_i - x_j) > 0$. If S shows a high positive value it means that there is an indicator of an increasing or upward trend, but if the S shows a very low negative value, it indicates a decreasing or downward trend. However, Khambhammettu [122] suggested that it is necessary to compute the probability associated with S and the sample size, to statistically quantify the significance of the trend.

Kendall [106] describes a normal approximation test that could be used for data sets with more than 10 values, providing that there are not many tied values within the data set. S calculation was described in the previous section. The mean E[S] and variance V [S] of the statistic S may be given using following equation:

$$E[S] = 0 \tag{3.45}$$

$$\mathrm{Var}[S] = \frac{n(n-1)(2n+5) - \sum_{p=1}^{q} t_p\left(t_p - 1\right)\left(2t_p + 5\right)}{18} \tag{3.46}$$

where:

n is the number of data points; q is the number of tied groups (a tied group is a set of sample data having the same value); and t_p is the number of ties for the pth value.

The second term represents an adjustment for tied data. t is the extent of any given tied data, $\sum_t$is the summation over all ties. In the case where the sample size $n > 10$, the standardized test statistic (Z_{MK}) is computed in Eq. (3.49) [126] as:

$$Z_{\mathrm{MK}} = \begin{cases} \frac{S-1}{\sqrt{\mathrm{Var}\ (S)}} & \text{if } S > 0 \\ 0 & \text{if } S = 0 \\ \frac{S+1}{\sqrt{\mathrm{Var}\ (S)}} & \text{if } S < 0 \end{cases} \tag{3.47}$$

Thus, in two-sided trend tests, H_0 is accepted if $|Z| \leq Z_{a/2}$ at a level of significance. Positive S indicates an "upward trend" while negative S indicates a "downward trend" [126]. The probability level of significance was considered at a level of 95 %. A positive Z_{MK} indicates an increasing trend. This means that the computed probability is more than the level of significance at a level of 95 %, whereas, a negative Z_{MK} indicates a decreasing trend. This means that the computed probability is also more than the level of significance at a level of 95 %. To test for either increasing or decreasing monotonic trends at p significance level, the null hypothesis is rejected if the absolute value of Z is greater than $Z_{1-p/2}$, where $Z_{1-p/2}$ is obtained from the standard normal cumulative distribution tables. In this work, the significance level of $p = 0.01$ and 0.05 are applied. Finally, if the computed probability is less than the level of significance, there is no trend [110].

3.12.4 Sequential Mann–Kendall Test

The sequential Mann–Kendall test Sneyers [127] is used to test an assumption about the beginning of the development of a trend within a sample ($x_1, \ldots, x_n$) of random variable x based on a rank series of progressive and retrograde rows of the sample. To see change of a trend with time, Sneyers [123] introduced sequential values, u(t) and u′(t), from the progressive analysis of the Mann–Kendall test. The Mann–Kendall test statistic u(t) is a value that indicates directions (or signs) and a statistical magnitude of the trend in a series. When the value of u(t) is significant at a 5 % significant level, it can be decided whether it is an increasing or a decreasing trend, depending on whether u(t) > 0 or u(t) < 0. A 1 % level of significance was also taken into consideration. Partial and short—period trends, and a change point or beginning point of a trend in climatic series were investigated by using time—series plot of the u(ti) and u′(ti) values. In order to obtain such a time- series plot, sequential values of the statistics u(t) and u′(ti) were computed from the progressive analysis of the Mann–Kendall test. The following steps are applied to calculate u(t) and u′(t):

First, original observations are replaced by their corresponding ranks y_i, which are arranged in ascending order. Then, for each term y_i, the number of terms y_i preceding ($i > j$) is calculated with ($y_i > y_j$). The values of x_j annual mean time series, ($j = 1,\ldots,n$) are compared with x_i, ($i = 1,\ldots, j - 1$). At each comparison, the number of cases $x_j > x_i$ is counted and denoted by n_j.

The test statistic t is then calculated by equation

$$t_j = \sum_{1}^{j} n_j \tag{3.48}$$

The distribution function of the test statistic t has a mean and a variance of the test statistic respectively, given as [128]:

$$E(t) = \frac{n(n-1)}{4} \tag{3.49}$$

$$\mathrm{Var}(t_j) = [j(j-1)(2j+5)/72] \tag{3.50}$$

The sequential values of the statistic u(t) are then computed as

$$\mathrm{U(t)} = \frac{t_j - E(t)}{\sqrt{\mathrm{Var}(t_j)}} \tag{3.51}$$

Finally, the values of u′(t) are similarly computed backward, starting from the end of the series using the same equation but in the inverse series of data.

In two-sided trend test, the null hypothesis is accepted at α significance level if $|\mathrm{U(t)}| \le \mathrm{U(t)}_{1=\alpha/2}$, where $\mathrm{U(t)}_{1=\alpha/2}$ is the critical value of standard normal distribution with a probability exceeding $\alpha/2$. Positive U(t) denotes a positive trend and negative U(t) denotes a negative trend. In this study, α was set at 0.05. The sequential version test enables detection of the approximate time of occurrence of a trend by locating the intersection point of the forward and backward curves of the test statistic. If the intersection point is significant at a $\alpha = 0.05$, we infer that the critical point of the analyzed time series occurs at that time [129, 130]. Hence, the Sequential Mann–Kendall test is considered an effective way of locating the beginning point of a trend [128].

3.13 Land Use Change Modeling

The first category of anthropogenic effects results from human actions on the ecosystems, such as deforestation and urbanization. These lead to changes in the ecosystem that magnify the consequences of heavy precipitation, converting this precipitation into floods of a greater severity than otherwise would have resulted IAEA/WMO [132]. Monitoring land use changes provides urban planners and decision makers with essential information about past and current states of development and the nature of changes that have occurred [122]. It is generally accepted that changes in land use patterns (e.g., expansion of settlements including road construction, deforestation, distinct practices in arable and grassland management) contribute to an increased frequency and severity of flood generation [123]. Development of land from forested cover to investigating pasture is known to increase the rate and the total volume of runoff during storm events [124, 125], Rowe [132]. Such land use changes results in higher flood peaks, and greater fluctuations in flow and water levels [126]. In this study, the data sources included were aerial photographs for the years of 1956, 1974, 1982, 1995, 2000, and 2004. The land use maps were digitized based on these available aerial photos. ArcGIS package was applied to detect the changes from 1956 to 2004. A geographic information system offered the possibility of structured data management and data access as well as efficient and objective analysis procedures. In light of these considerations, the present work aims to propose a statistical analysis for the assessment of the land use change over the time and find any probable effect on the flow propagation of the Ubaye River.

3.14 Chapter Summary

This chapter has described the methodologies applied in this research and introduced all the procedures in detail. For quality control of the DEM, different interpolation techniques were introduced and it was also explained how the results might be analyzed both statistically and hydrologically. Then, it was explained how the spatial and temporal input data such as discharge, RAS-Geometry data and cross sections were prepared for hydrodynamic simulation purposes. In the next step, the calibration and modeling approaches for flood hazard assessment, the basic theory for hydraulic and hydrologic models (HEC-RAS and SOBEK), were described. Both models generate output data such as water depth, velocity, and inundation in the form of map-series, and the rationale behind these maps was also explained in this chapter. Continuing on, the designed scenarios and frequency analysis and the aims behind them were described, which is necessary for any future flood management in the case study. In the next part, the trend testing approaches for hydrometeorological variables such as precipitation, temperature, and stream flow were formulated for the time series of 1961–2009. The trend analysis of hydrological and meteorological series is important and even more relevant when considering the regional effects of global climate change. Trend analysis provides an idea about climatic variability in the watershed. Kendall's method, as a strong method, which is suggested by many investigators, introduced to check the significance of trends in hydroclimatological parameters. Finally, the process for preparing and digitizing the land use maps based on available aerial photos was described shortly. The statistical analysis was applied to compare the changes from the past until the present. The next chapter provides a background on the study area and characteristic of Ubaye River.

References

1. Vaze J, Teng J, Spencer G (2010) Impact of DEM accuracy and resolution on topographic indices. Environ Modeling Softw 25:1086–1098
2. Alkema D, Hiddelkoop H (2005) The influence of floodplain compartmentalization on flood risk within the Rhine-Meuse Delta. Nat Hazards 36:125–145
3. Alkema D (2007) Simulating floods, on the application of a 2D-hydraulic model for flood hazard and risk assessment, Ph.D. Dissertation, International Institute for Geo-information Science and Earth Observation (ITC), Twente University, Enschede, The Netherlands
4. Burrough PA, McDonnell RA (1998) Principals of geographical information systems. Oxford University Press, New York, p 333
5. Abd Rahman M and Alkema D (2006) Digital Surface Model (DSM) Construction and Flood Hazard Simulation for Development Plans in Naga City, Philippines. GIS Development Malaysia, July-September, Vol.1, Issue 3
6. Johnston K, Ver-Hoef JM, Krivoruchko K, Lucas N (2001) Using ArcGIS geostatistical analyst. ESRI, Redlands, p 300
7. Peralvo M (2008) GIS in water resources; influence of DEM interpolation methods in drainage analysis, Available online at: http://en.scientificcommons.org/manuel_peralvo (Verified 28 Nov 2011)
8. Wise SM (2000) Assessing the quality for hydrological applications of digital elevation models derived from contours. Hydrol Process 14:1909–1929
9. Wise SM (2007) Effect of differing DEM creation methods on the results from a hydrological model. Comput Geosci 33:1351–1365
10. Rahman AA, Piloul M (2008) Spatial data modelling for 3D GIS. Springer, New York
11. Bhattacharya N (2010) Flood risk assessment in Barcelonnette, France, M.Sc. Thesis, International Institute for Geo-information Science and Earth Observation(ITC), Twente University, Enschede, The Netherlands

12. Franke R (1982) Smooth interpolation of scattered data by local thin plate splines. Comput Math App 8(4):273–281
13. Mitas L, Mitasova H (1988) General variational approach to the interpolation problem. Comput Math App 16(12):983–992
14. ArcGIS 9.3-online Help, Available online at: http://webhelp.esri.com (Verified 28 Nov 2011)
15. Philip GM, Watson DF (1982) A precise method for determining contoured surfaces. Australian Pet Explor Assoc J 22:205–212
16. Watson DF, Philip GM (1985) A refinement of inverse distance weighted interpolation. Geoprocessing 2:315–327
17. Environmental Systems Research Institute (ESRI) (1996) ArcView spatial analyst. Environmental Systems Research Institute, Inc., Redlands, p 147
18. Wechsler S (1999) Results of the DEM user survey. Available online at: http://web.syr.edu/~sperlit/survey (Verified 28 Nov 2011)
19. Hutchinson MF (1988) Calculation of hydrologically sound digital elevation models. In: Proceedings of the third international symposium on spatial data handling, Aug 17–19, Sydney. International Geographical Union, Columbus, Ohio, pp 117–133
20. Hutchinson MF (1989) A new procedure for gridding elevation and stream line data with automatic removal of spurious pits. J Hydrol 106:211–232
21. Wahba G (1990) Spline models for observational data, CNMSNSF Regional conference series in applied mathematics 59. SIAM, Philadelphia
22. McBratney AB, Webster R (1986) Choosing functions for semivariograms of soil properties and fitting them to sampling estimates. Soil Sci 37:617–639
23. Press WH, Teukolsky SA, Vetterling WT, Flannery BP (1988) Numerical Recipes in C, The Art of Scientific Computing. Cambridge University Press, New York
24. Levenberg K (1944) A method for the solution of certain non-linear problems in least squares. Q Appl Math 2:164–168
25. Gundogdu KS, Guney I (2007) Spatial analyses of groundwater levels using universal kriging. J Earth Syst Sci 116:49–55
26. Fisher PF (1998) Improved modelling of elevation error with geostatistics. GeoInformatica 2:215–233
27. Monckton C (1994) Chapter 14: An investigation into the spatial structure of error in digital elevation data, innovations in GIS, vol I Worboys M (ed). Taylor and Francis, Bristol, p 267
28. Wise SM (1998) The effect of GIS interpolation errors on the use of digital elevation models in geomorphology. In: Lowe SN, Richards KS, Chandler JH (eds) Landform monitoring, modelling and analysis. Wiley, New York, pp 139–164
29. Gong J, Li Z, Zhu Q, Sui H, Zhou Y (2000) Effects of various factors on the accuracy of DEMs: an intensive experimental investigation. Photogrammetric Eng Remote Sens 66:1113–1117
30. Chang K, Tsai B (1991) The effect of DEM resolution on slope and aspect mapping. Cartography Geog Inf Syst 18:69–77
31. Carter JR (1992) The effect of data precision on the calculation of slope and aspect using gridded DEMs. Cartographica 29:22–34
32. Bolstad PV, Stowe T (1994) An evaluation of DEM accuracy: elevation, slope and aspect. Photogrammetric Eng Remote Sens 60:1327–1332
33. Albani M, Klinkenberg B (2003) A spatial filter for the removal of striping artifacts in digital elevation models. Photogrammetric Eng Remote Sens 63:755–766
34. Brown DG, Bara TJ (1994) Recognition and reduction of systematic error in elevation and derivative surfaces from 7.5 minute DEMs. Photogrammetric Eng Remote Sens 60:189–194
35. Guth PL (1999) Contour line ghosts in USGS level 2 DEMs. Photogrammetric Eng Remote Sens 65:289–296
36. Wood J, Fisher PF (1993) Assessing interpolation accuracy in elevation models. IEEE Comput Graphics Appl 13:48–56

37. Florinsky IV, Kuryakova GA (2000) Determination of grid size for digital terrain modeling in landscape investigations—exemplified by soil moisture distribution at a mircoscale. Int J Geo Info Sci 14(8):815–832
38. Wechsler SP, Kroll C (2006) Quantifying DEM uncertainty and its effect on topographic parameters. Photogrammetric Eng Remote Sens 72:1081–1090
39. Blom G (1958) Statistical estimates and transformed beta variables. Wiley, New York, pp 143–146
40. Moran PAP (1950) Notes on continuous stochastic phenomena. Biometrika 37(1):17–23
41. Fernando T, Rangel LVB, Field, R, Alexandre J, Diniz-Filho F (2011) SAM Spatial Analysis in Macroecolog, SAM Tutorial, SAM workshop, IBS meeting, Crete 07 Jan 2011
42. Maidment DR, Olivera F, Calver A, Eatherral A, Fraczek W (1996) A unit hydrograph derived from a spatially distributed velocity field. Hydrol Processes 10:831–844
43. Tribe A (1992) Automated recognition of valley heads from digial elevation models. Earth Surf Processes Landforms 16(1):33–49
44. Garbrecht J, Martz LW (1995) Advances in automated landscape analysis. In: Espey WH and Combs PG (eds) Proceedings of the first international conference on water resources engineering, vol 1. American Society of Engineers, San Antonio, Aug 14–18, pp 844–848
45. Garbrecht J, Martz LW (1997) The assignment of drainage direction over flat surfaces in raster digital elevation models. J Hydrol 193:204–213
46. Garbrecht J, Martz LW (1999) TOPAZ: An automated digital landscape analysis tool for topographic evaluation, drainage identification, watershed segmentation and sub catchment parameterization; TOPAZ Overview. U.S. Department of Agriculture, Agricultural Research Service, Grazing lands Research Laboratory, El Reno, Oklahoma, USDA, ARS Publication No. GRL 99-1, p 26
47. Hutchinson MF, Dowling TI (1991) A continental hydrological assessment of a new grid-based digital elevation model of Australia. Hydrol Process 5:45–58
48. Horritt MS, Bates PD (2002) Evaluation of 1D and 2D numerical methods for predicting river flood inundation. J Hydrol 268:87–99
49. Chow VT (1959) Open channel hydraulics. McGraw-Hill, New York
50. Visser M, Duits M, Van der Lee W (2001) An automatic calibration method for hydraulic models. NCR-days 2001:48–49
51. Muthukishnan S, Harbor J, Lim K, Engel B (2006) Calibration of a simple rainfall-runoff model for long-term hydrological impact evaluation. Urban Reg Inf Syst Assoc J (18/2):35–42
52. Wijbenga JHA, Duits MT, van Noortwijk JM (1998) Parameter optimisation for two dimensional flow modelling. In: Babovic V, Larsen LC (eds) Hydroinformatics 98. Rotterdam, Brookfield, pp 1037–1042
53. Nelder JA, Mead R (1965) A simplex method for function minimisation. Comput J 7:308–313
54. Wisser D, Fekete BM, Vörösmarty CJ, Schumann AH (2010) Reconstructing 20th 762 century global hydrography: a contribution to the Global Terrestrial 763 Network-Hydrology (GTN-H). Hydrology and Earth System Sciences 14(1): 764 1–24
55. Van der Veen R, Pakes U, Schutte L in voorbereiding. Bouw en calibratie Sobek Rijn, versie 2000.1 t/m 2000.3. RIZA rapport, Rijkswaterstaat/RIZA, Arnhem (In Dutch)
56. Boughton WC, Droop O (2003) Continuous simulation for design flood estimation—a review. Environ Modelling Softw 18(4):309–318
57. Alemseged TH, Rientjes THM (2007). Uncertainty issues in hydrodynamic flood modeling. In: Proceedings of the 5th international symposium on spatial data quality SDQ 2007, Modelling qualities in space and time, International Institute for Geo-information Science and Earth Observation(ITC), Twente University, Enschede, The Netherlands, 13–15 June 2007
58. Fread DL (1992) Flow routing. In: Maidment DR (ed) Handbook of hydrology. McGraw-Hill, Inc. New York, pp 10.1–10.36
59. Stelling GS, Verwey A (2005) Numerical flood simulation, encyclopaedia of hydrological sciences. Wiley, UK

60. Pudasaini SP, Hutter K (2007) Avalanche dynamics: dynamics of rapid flows of dense granular avalanches. Springer, Berlin. ISBN 3-540-32686-3
61. Takahashi T (2007) Debris flow: mechanics, prediction and countermeasures. Taylor and Francis, London, p 448
62. Pudasaini SP, Domnik B (2009) Energy considerations in accelerating rapid shear granular flows. Nonlinear Processes Geophys 16:399–407
63. Vaughan CL, O'Malley MJ (2005) Froude and the contribution of naval architecture to our understanding of bipedal locomotion. Gait Posture 21(3):350–362 (In French)
64. HEC-RAS manual (2009) The Hydrologic Engineering Center (HEC), Institute for Water Resources, US ArmyCorps of Engineer, Available online at: http://www.hec.usace.army.mil/software/hec-ras/ (Verified 28 Nov 2011)
65. Pelinovsky E, Kozlov S, Vinogradov S (2002) Numerical simulation of spring floods in Nizhny Novgorod region (RUSSIA). In: PACON proceedings (symposium on humanity and the world ocean: in terdependence at the dawn of the new millennium, June 23–25, 1999, Moscow, pp 391–399
66. Bin Usamah M, Alkema D (2006) Simulation scenario floods for hazard assessment on the lower Bicol floodplain, the Philippines. In: Advances in Geosciences: vol 4. Hydrological science: Papers from the proceedings of Asia Oceania Geosciences Society AOGS 3rd annual meeting: 10–14 July 2006, pp 69–74
67. SOBEK User's Manual (2003) developed by Delft Hydraulics, The Netherlands, Available online at:http://delftsoftware.wldelft.nl(Verified 28 Nov 2011)
68. Van Westen C, Alkema D, Vargas R, Quan Luna B (2010) Training materials: GIS for multi-hazard mountain risk assessment, Intensive Risks Course, mountain disaster risk management, 19–25 June 2010, Barcelonnette and Avignonet, France
69. Diplas P, Sutherland A (1988) Sampling techniques for gravel sized sediments. J Hydraulic Eng 114:481–501
70. COM (2004) 472 final, Brussels, 12.07.2004, Commission of the European communities, Communication From the commission to the council, The European Parliament, The European economic and social committee and the committee of the regions, Flood risk management, Flood prevention, protection and mitigation
71. Abbott MB, Bathurst JC, Cunge JA, Connell PE, Rasmussen J (1986) An introduction to the European Hydrological System-Systeme Hydrologique Eropeen, 'SHE', 1: History and Philosophy of a physically based, distributed modelling system. J Hydrol 87:45–59
72. Rohdenburg H, Diekkruger B, Bork HR (1986) Deterministic hydrological site and catchment models for the analysis of agroecosystems. Catena 13:119–137
73. Kenward T, Lettenmaier DP, Wood EF, Fielding E (2000) Effects of digital elevation model accuracy on hydrologic predictions. Remote Sens Environ 74(3):432–444
74. Wolock DM, McCabe GJ (2000) Differences in topographic characteristics computed from 100- and 1000-m resolution digital elevation model data. Hydrol Process 14:987–1002
75. Jenson SK (1991) Applications of hydrological information automatically extracted from digital elevation models. Hydrol Process 5:31–44
76. Jenson SK, Domingue JO (1988) Extracting topographic structure from digital elevation data for geographic information system analysis. Photogrammetric Eng Remote Sens 54:1593–1600
77. Hutchinson MF (1996) A locally adaptive approach to the interpolation of digital elevation models. In: NCGIA (ed) Proceedings of the third international conference integrating GIS and environmental modeling, Santa Fe, 21–25 Jan, 1996. University of California, Santa Barbara, National Center for Geographic Information and Analysis: CD-ROM and Available online at: http://www.ncgia.ucsb.edu/conf/SANTA_FE_CD-ROM/main.html (Verified 28 Nov 2011)
78. Hutchinson MF, Gallant JC (1999) Representation of terrain, Geographical Information Systems: Principles, Technical Issues, Management Issues and Applications. Second Edition. Edited by Longley PA., Goodchild MF, Maguire DJ and Rhind DW. Wiley, New York, Chapter 9, pp 105–124.

79. Schoorl JM, Sonneveld MPW, Veldkamp A (2000) Three-dimensional landscape process modelling: the effect of DEM resolution. Earth Surf Proc Land 25:1025–1035
80. Warren SD, Hohmann MG, Auerswald K, Mitasova H (2004) An elevation of methods to determine slope using digital elevation data. Catena 58:215–233
81. Wilson JP, Gallant JC (2000) Differences in topographic characteristics computed from 100- and 1000-m resolution digital elevation model data. Hydrol process 14:987–1002
82. Van Baars S (2005) The horizontal failure mechanism of the Wilnis peat dyke. Geotechnique 55(4):319–323
83. Apel H, Merz B, Thieken AH (2009) Influence of dyke breaches on flood frequency estimation. Comput Geosci 35(5):907–923
84. Apel H, Thieken A, Merz B, Blöschl G (2004) Flood risk assessment and associated uncertainty. Nat Hazards Earth Syst Sci 4:295–308
85. Möllmann A, Vermeer PA (2007) Reliability analysis of a dike failure, XVII European Young Geotechnical Engineers' Conference, Presentation within the framework of "PC River—Reliability and risk analysis in river flood protection under consideration of geotechnical, hydrological and hydraulic factors", June, 18th, 2007, Ancona
86. Dalrymple RA, Herault A (2009) Levee Braching with GPU-SPHysics code, 4th international SPHERIC workshop, Nantes, France, 27–29 May 2009
87. Kamrath P, Disse M, Hammer M, Köngeter J (2006) Assessment of discharge through a dike breach and simulation of flood wave propagation. Nat Hazards 387(1–2):63–78
88. Jothityangkoon C, Sivapalan M (2003) Towards estimation of extreme floods; examination of the roles of runoff process changes and floodplain flows. J Hydrol 281(3):206–229
89. Woltemade CJ, Potter KW (1994) A watershed modelling analysis of fluvial geomorphologic influences on flood peak attenuation. Water Resour Res 30(6):1933–1942
90. Ramesh A, Glade T, Parodi G, Alkema D (2011) Dyke failure scenarios in Ubaye River, France using hydraulic modelling, European Geosciences Union (EGU), Vienna, Austria, 03—08 April 2011
91. Horritt MS, Bates PD (2001) Effects of spatial resolution on a raster based model of flood flow. J Hydrol 253:239e249
92. Parodi G (2009) Frequency analysis for hydrologic modelling, International Institute for Geo-information Science and Earth Observation (ITC), Twente University, Enschede, The Netherlands
93. Jain S, Lall U (2000) Magnitude and timing of annual maximum floods: trends and large scale climatic associations for the Blacksmith Fork River, Utah. Water Resour Res 36:3641–3651
94. Milly PCD, Wetherald RT, Dunne KA, Delworth TL (2002) Increasing risk of great floods in a changing climate. Nature 415:514–517
95. Kingston DG, Lawler DM, McGregor GR (2006) Linkages between atmospheric circulation, climate and stream flow in the northern North Atlantic: research prospects. Prog Phys Geogr 30:143–174
96. Helms M, Büchele B, Merkel U, Ihringer J (2002) Statistical analysis of the flood situation and assessment of the impact of diking measures along the Elbe (Labe) river. J Hydrol 267:94–114
97. Lammersen R, Engel H, Van de Langemheen W, Buiteveld H (2002) Impact of river training and retention measures on flood peaks along the Rhine. J Hydrol 267:115–124
98. Drogue G, Pfister L, Leviandier T, El Idrissi A, Iffly JF, Matgen P, Humbert J, Hoffmann L (2004) Simulating the spatio-temporal variability of stream flow response to climate change scenarios in a mesoscale basin. J Hydrol 293(1–4):255–269
99. Stedinger JR, Vogel RM, Foufoula-Georgiou E (1993) Frequency analysis of extreme events. In: Maidment DR (ed) Handbook of hydrology. McGraw-Hill, New York, pp 18.1–18.66
100. Hosking JRM, Wallis JR (1997) Regional frequency analysis, an approach based on L-Moments. Cambridge University Press, Cambridge, p 240

101. Hirsch RM, Slack JR, Smith RA (1982) Techniques of trend analysis for monthly water quality data. Water Resour Res 18(1):107–121
102. Helsel DR, Hirsh RM (1991) Hydrologic analysis and interpretation, statistical methods in water resources, techniques of water-resources investigations of the USGS book 4, Chapter A3. Available online at: http://pubs.usgs.gov/twri/twri4a3/ (Verified 28 Nov 2011)
103. Mann HB (1945) Nonparametric tests against trend. Econometrica 13:245–259
104. Kendall MG (1975) Rank correlation methods. Charles Griffin, London, p 202
105. Burn DH, Hag Elnur MA (2002) Detection of hydrologic trend and variability. J Hydrol 255:107–122
106. Yue S, Pilon P, Phinney B, Cavadias G (2002) The influence of autocorrelation on the ability to detect trend in hydrological series. Hydrol Process 16:1807–1829
107. Partal T, Kahya E (2006) Trend analysis in Turkesh precipitation data. Hydrol Process 20:2011–2026
108. Nasri M, Modarres R (2009) Dry spell trend analysis of Isfahan Province, Iran. Int J Climatol 29:1430–1438
109. Buish TA (1982) Some methods for testing the homogeneity of rainfall records. J Hydrol 58:11–27
110. Wijngaard JB, Klein Tank MG, Können GP (2003) Homogeneity of 20th century European daily temperature and precipitation series. Int J Climatol 23:679–692
111. Lazaro R, Rodrigo FS, Gutierrez L, Domingo F, Puigdefabregas J (2001) Analysis of a 30-year rainfall record (1967–1997) in semiarid SE Spain for implications on vegetation. J Arid Environ 48:373–395
112. Burn DH (1994) Hydrological effects of climatic change in west-central Canada. J Hydrol 160:53–70
113. Burn DH, Cunderlik JM, Pietroniro A (2004a) Hydrological trends and variability in the Liard River Basin. Hydrol Sci J 49(1):53–67
114. Burn DH, Abdul Aziz OI, Pietroniro A (2004b) A comparison of trends in hydrometeorological variables for two watersheds in the Mackenzie River Basin. Canadian Water Resour J 29(4):283–298
115. Zhu T, Jenkins M, Lund J (2005) Estimated impacts of climate warming on california water availability under twelve future climate scenarios. J Am Water Resour Assoc 41(5):1027–1038
116. Gilbert RO (1987) Statistical methods for environmental pollution monitoring. Van Nostrand Reinhold, New York
117. Libiseller C, Grimvall A (2002) Performance of partial Mann-Kendall test for trend detection in the presence of covariates. Environmetrics 13:71–84
118. Dietz EJ, Kileen A (1981) A nonparametric multivariate test for monotone trend with pharmaceutical applications. J Am Stat Assoc 76:169–174
119. Lettenmaier DP (1988) Multivariate nonparametric tests for trend in water quality. Water Resour Bull 24(3):505–512
120. Khambhammettu P (2005) Mann-Kendall analysis, HydroGeoLogic Inc.—U-1 Annual groundwater monitoring report—former fort ord, California
121. Sneyers R (1990) On the statistical analysis of series of observations. Tech Note 143, WMO-No. 415, p 192
122. Huth R (1999) Statistical downscaling in central Europe: evaluation of methods and potential predictors. Climate Res 13:91–101
123. Hirsch RM, Slack JR (1984) A nonparametric trend test for seasonal data with serial dependence. Water Resour Res 20:727–732
124. Douglas EM, Vogel RM, Kroll CN (2000) Trends in floods and low flows in the united states: impact of spatial correlation. J Hydrol 240:90–105
125. Tian F, Yang Y, Han S (2009) Using runoff slope-break to determine dominate factors of runoff decline in Hutuo River Basin, North China. Water Sci Technol 60(8):2135–2144

126. Moraes JM, Pellegrino HQ, Ballester MV, Martinelli LA, Victoria R, Krusche AV (1998) Trends in hydrological parameters of southern Brazilian watershed and its relation to human induced changes. Water Resour Manage 12:295–311
127. Gerstengarbe FW, Werner PC (1999) Estimation of the beginning and end of recurrent events within a climate regime. Climate Res 11:97–107
128. Musaoglu N, Coskun M, Kocabas V (2005) Land use change analysis of Beykoz-Istanbul by means of satellite images and GIS. Water Sci Technol 51(11):245–251
129. IAEA/WMO (2006) Global Network of Isotopes in Precipitation. The GNIP Database. (http://isohis.iaea.org), accessed August 2010
130. Wahren A, Feger K-H, Schẅarzel K, M̈unch A (2009) Land-use effects on flood generation—considering soil hydraulic measurements in modelling. Adv Geosci 21:99–107
131. Selby MJ (1972) The relationship between land use and erosion in the Central North Island, New Zealand. J Hydrol 11:73–85
132. Jackson RJ (1973) Catchments hydrology and some of its problems proceedings of soil and plant water symposium, Palmerston North, N.Z. DSIR Inf Series, 96:73–80
133. Rowe LK, Pearce AJ (1994) Hydrology and related changes after harvesting native forest catchment and establishing Pinus Radias Plantations. Part 2. The native forest water balance and changes in streamflow after harvesting, Hydrological processes 8:281–297
134. Hamilton D (2001) Waikato project watershed, contributor—hydrology background technical Paper, Unpublished report prepared for Environment Waikato, May 2001
135. Skidmore AK, Naimi B, Nieuwenhuis W, Toxopeus B (2010) Spatial autocorrelation, NRM_ module 13 _ spatial modelling, international institute for geo-information science and earth observation(ITC), Twente University, Enschede, The Netherlands
136. Akkala A, Devabhaktuni V, Kumar A (2010) Interpolation techniques and associated software for environmental data. Environ Prog Sustainable Eng 29:134–141
137. Haile AT (2005) Effects of LIDAR DEM Resolution in flood modeling: a model sensitivity study for the city of Tegucigalpa, Honduras. ISPRS WG III/3, III/4, V/3 Workshop "Laser scanning 2005", International Institute for Geo-information Science and Earth Observation(ITC), Twente University, Enschede, The Netherlands, Sept 12–14, 2005
138. Dutta D, Herath S, Musiake K (2000) Flood inundation simulation in a river basin using a physically based distributed hydrologic model. Hydrol Processes 14(3):497–519
139. Ruji EM (2007) Floodplain inundation simulation using 2D hydrodynamic modeling approach. M.Sc. Thesis, International Institute for Geo-information Science and Earth Observation (ITC), Twente University, Enschede, The Netherlands

Chapter 4
Study Area

4.1 Introduction

This chapter provides a description of the Barcelonnette area and the Ubaye River. The climatic conditions, the type of land use, geology, and soil characteristics are also described in this chapter, in order to clarify the study area and to better understand the problem statements.

4.2 Barcelonnette Area

This study focuses on the floodplain of the Ubaye River in the Barcelonnette (Barciloneta/Barcilona in Occitan) area, which is located in the Alpes-de-Haute-Provence, Southeast of France. The area is divided into four subbasins of Barcelonnette, Saint-Pons, Les Thuiles, and Uvernet-Fours, which are parts of the whole Ubaye catchment. The reason for choosing this area as a case study was because of the available hydrologic data at the Barcelonnette station, which were defined as boundary conditions of hydrodynamic models as inlet input data. For the upper part of the whole catchment there is not enough data, such as discharge data. The floodplain extends over an area of 93 km^2 with a length of 14 km and a maximum width of 9 km. It lies at an elevation of 1132 m (3717 ft) between latitude of 44° 19' 30"N to 44° 27' 00"N and longitude of 6° 33' 00"E to 6° 40' 00"E (Fig. 4.1).

A. Ramesh, *Response of Flood Events to Land Use and Climate Change*, Springer Theses, DOI: 10.1007/978-94-007-5527-7_4,

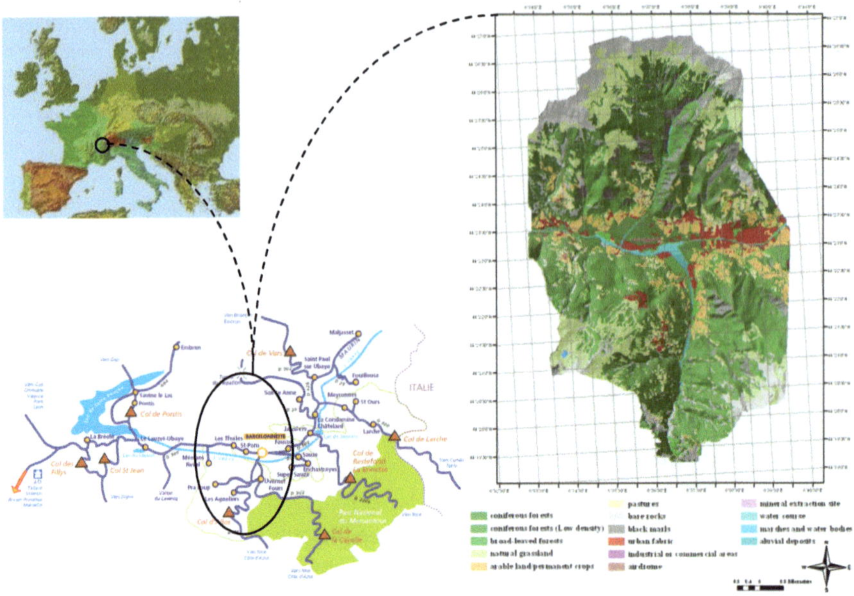

Fig. 4.1 Location of case study (Barcelonnette area) Alpes-de-Haute-Provence, France

The Ubaye Valley is representative of climatic, lithological, geomorphological, and land use conditions observed in the South French Alps [1]. The study area is located in the dry intra-Alpine zone, characterized by a mountainous Mediterranean climate with a high interannual rainfall variability (400–735 mm over the period of 1928–2005), a mean annual temperature of a ca. 7.5°, and the presence of a snow pack on the upper slopes for 4–6 months out of the year. Local summer rainstorms can be intense, sometimes yielding more than 40 mm.h^{-1} [2].

4.3 Geographical Characteristics of the Ubaye River in the Whole Watershed

According to the data bank from 1903 to 1990, Parde [3] described the Ubaye River as a fifth-order tributary of the middle Durance for the whole catchment. He mentioned that it is a steep river (13 %) with a mean annual discharge of 10.1 (m^3.s^{-1}) (Table 4.1).

Table 4.1 Geographical characteristics of the Ubaye River [3, 4]

Stream order	5
Length (km)	82
Watershed area (km^2)	970
Watershed geology	Flysch and black marls
Mean annual discharge at Barcelonnette ($m^3.s^{-1}$)	10.9
1 in 1.5 year discharge[a] ($m^3.s^{-1}$)	62(B.H.)
1 in 10 year discharge ($m^3.s^{-1}$) (B.H.)[a]	130
Hydrological regime (Parde 1995)	Snow regime influenced by Mediterranean climate
Altitudes (m)	900–1210
Slope (m per km)	12.92
Stream power ($w.m^{-2}$)	30–40 [b] å 400–600 [c]
Sinuosity rate	1.06

[a] Gumbel adjustment
[b] In the braiding segment, In the downstream gorges
[c] B.H. HYDRO data bank/Barcelonntte station (1903–1990)

According to Parde [3] and [4], the Ubaye River is characterized by a transitional snow regime. The maximum rainfall in the summer combines with a secondary peak in autumn. The whole area is located at a high altitude (900–1200 m). The alluvial corridor, oriented west–east, becomes progressively narrower downstream, as the last 6 km before the confluence is in gorges. This steep corridor is controlled by alluvial fans, many of which are no longer active. In open parts of the basin near Barcelonnette, the river has an active channel of roughly 300–400 m in width. In narrower sectors, such as the Thuiles-Meolans segment, the channel meanders. This amplitude becomes increasingly constrained by the valley's narrowness downstream, while the fluvial pattern becomes less sinuous and straight [3–5]. Figure 4.2 shows the Ubaye River with a floodplain area as a selected case study.

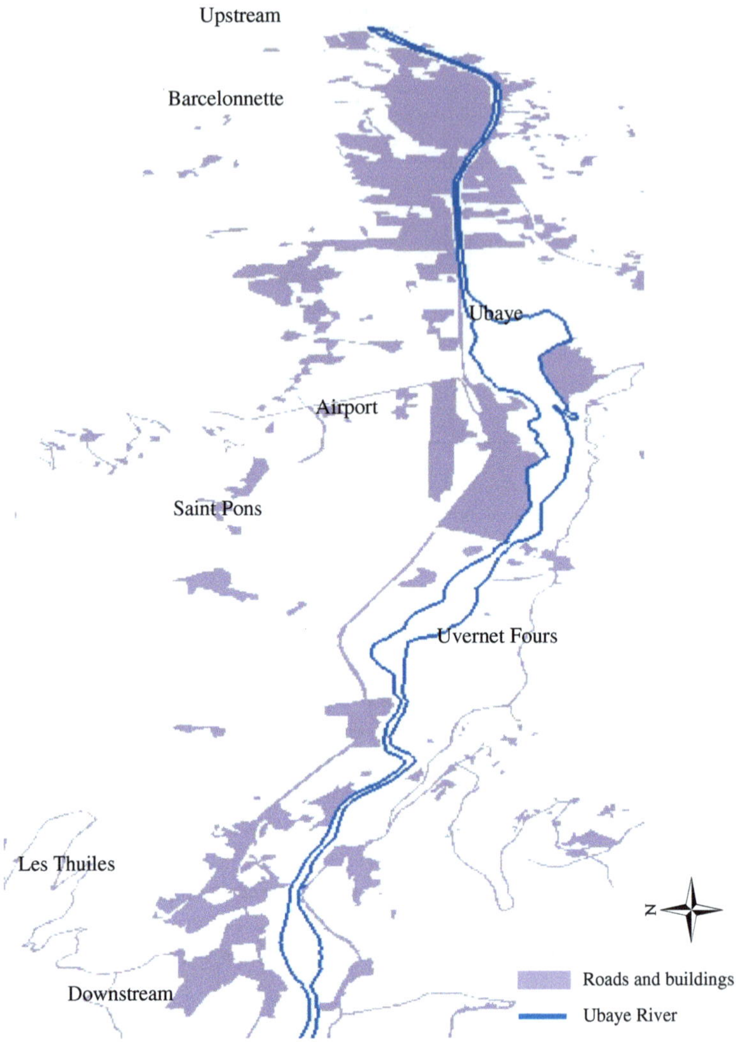

Fig. 4.2 Ubaye River with flood plain area

4.4 Climate

The Barcelonnette Basin is affected by a Mediterranean marine influence, moderated by a mild continental mountain climate. The orographic influence causes precipitation to increase with an elevation typically in the form of an "attenuated Mediterranean regime" [6].

The basin has a dry and mountainous Mediterranean climate (OMIV) with strong interannual rainfall variability. The measurement station is located in the

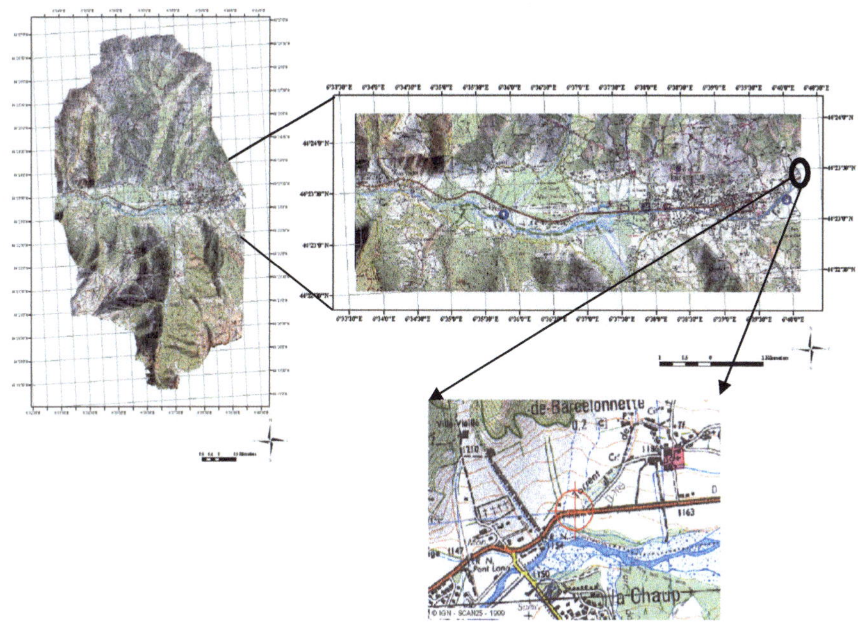

Fig. 4.3 The location of measurement station in the area

Barcelonnette city at an elevation of 1155 m. The station was created on January 1923. Service access is under the Meteo France Organization: (http://climatheque.meteo.fr). Figure 4.3 shows the location of the meteorological station in the case study.

The Barcelonnette area forms an elongated shape and villages are usually found along channels, rivers, or road sites that promote attenuated settlement forms [7]. This elongated structure makes it highly dependent on structural measures, such as dykes and levees, to protect against flooding. The Ubaye Valley is representative of climatic, lithological, geomorphologic, and land use conditions observed in the south French Alps, and is highly affected by flood hazards [1]. The Barcelonnette area is a tourist center, hosting many ski lodges. The whole Barcelonnette basin is known for its numerous slope (the slope gradients range from 10 to 20°) instabilities and other geomorphological hazards [6, 8] as well as highly active flood hazards which directly and indirectly affect land use and people's lives in the area.

4.4.1 Precipitation

Observed precipitation data in the Barcelonnette station from 1961 to 2009 show that the highest cumulative annual precipitation of 1027.6, 1191.8, and 1781.6 mm was recorded in 1963, 1976, and 1977, respectively (Fig. 4.4). Conversely, the lowest amount of precipitation occurred in 2007 with an amount of 384 mm.

Fig. 4.4 Annual precipitation between 1961 and 2009

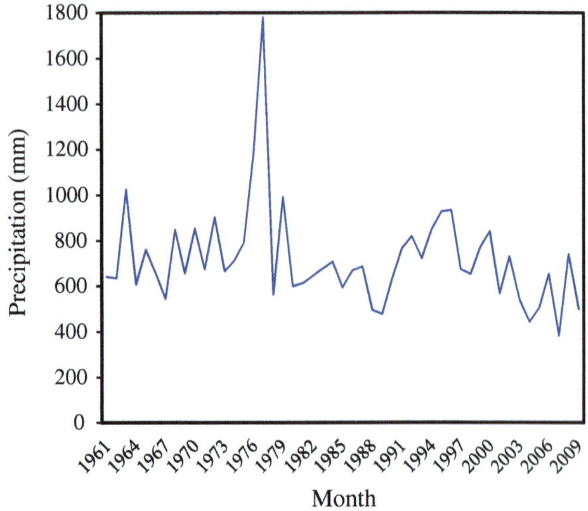

Fig. 4.5 Monthly precipitation between 1961 and 2009

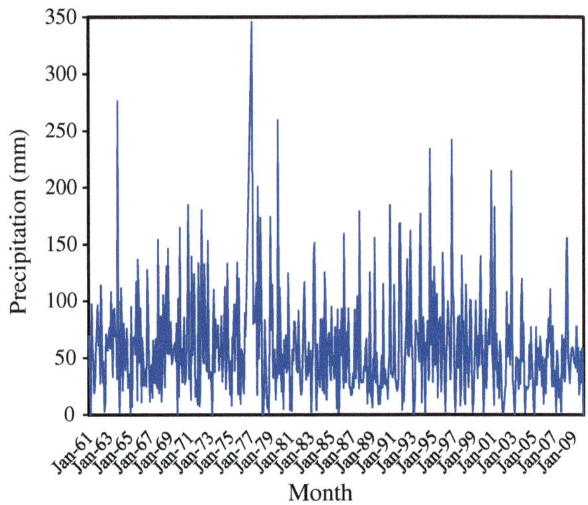

In 1963, 1976, and 1977, the maximum monthly precipitation was 277, 242.3, and 346.1 mm, respectively (recorded in November, December, and March). Between 1961 and 2009, the maximum rainfall was in spring and autumn, with the first peak occurring in May and June and the second peak occurring in October and November (Figs. 4.5 and 4.6).

For the recent flood event of 2008, the amount of precipitation recorded in May was 156 mm; however, almost the same value was recorded for 1983 but no flood was reported for this year. It is obviously clear that there was a triggering factor for occurrence of flood in 2008.

Although a similar precipitation value to the 2008 flood event was recorded for 1983, no flood was reported.

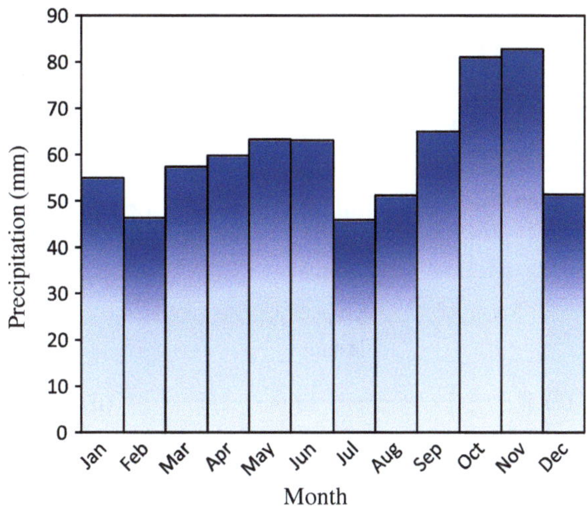

Fig. 4.6 Average monthly precipitation between 1961 to 2009

4.4.2 Temperature

Based on the observed data in the Barcelonnette station from 1961 to 2009, the maximum recorded temperature in August 2003 was 29.37 °C while the minimum temperature in January 1981 was 13.59 °C (Fig. 4.7a–f).

It can be seen through a comparison that the precipitation and discharge in August 2003 had a maximum value of 48.4 and a minimum value of 5.64. In January 1981, the precipitation and discharge were recorded at 33.3 and 5.21 mm.

The monthly analysis shows that the highest temperature was during May to September, but July and August always show the highest temperatures, more than 20 °C (Fig. 4.8).

4.4.3 Snow Depth

The snow depth observation was limited from only 1995 to 2005, which was recorded at Restofond at an elevation of 2720 m, with the slope facing to the north. According to Weber's [6] investigation, some factors such as air temperature, wind direction, and wind velocity affect the depth of snow in this area. An average of 1.3 m snow depth was recorded in the Barcelonnette area in the valley region, while this rate increased to 8 m in the skiing area at an elevation of 2500–3000 m height on the shady southern slope [6]. According to the literature, the duration of snow cover lasts at least 4–5 months per year (Fig. 4.9). Bhattacharya [9] conducted some interviews with local people and they reported that there had been neither any decline in the amount of snowfall nor had there been more frequent or wetter winters. Drier summers are a possible indication of climate change in

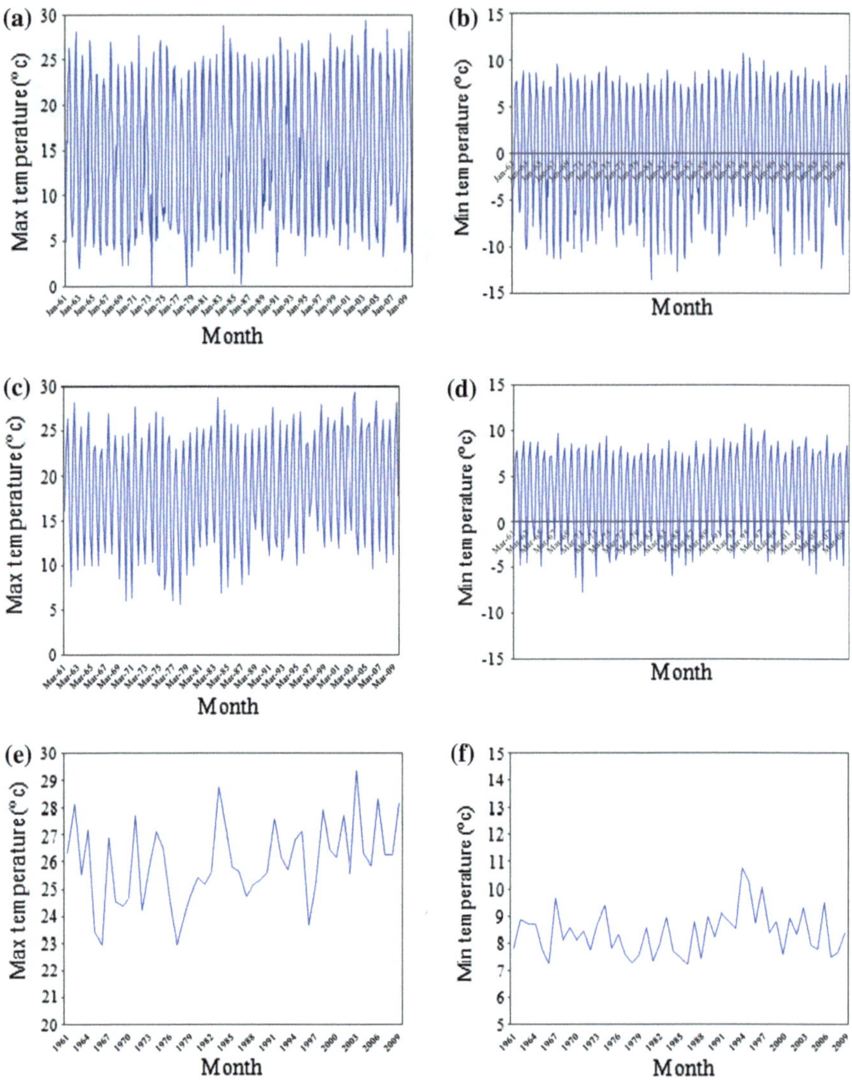

Fig. 4.7 **a** Monthly maximum temperature. **b** Monthly minimum temperature. **c** Seasonally maximum temperature. **d** Seasonally minimum temperature. **e** Annual maximum temperature. **f** Annual minimum temperature

the area. Based on available data, the maximum snow depth was recorded in September 2001.

In addition to the previously mentioned influence of the Mediterranean climate and the mild continental and mountain climate, the area is also affected by a warm dry wind coming from the south, called "Sirocco", which causes high snowmelt in spring and then high discharge in the Ubaye River [6].

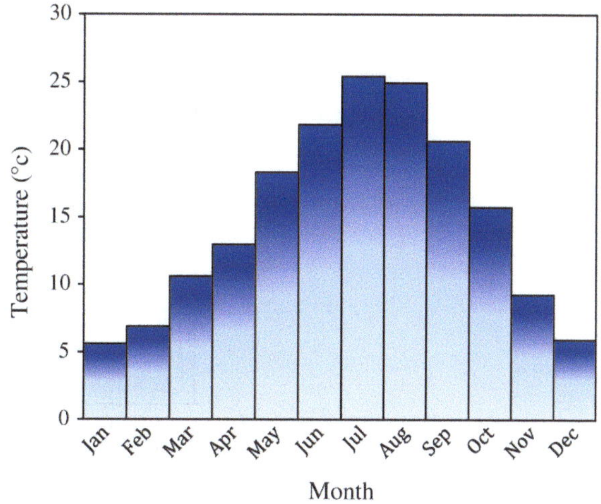

Fig. 4.8 Average monthly temperature between 1961 and 2009

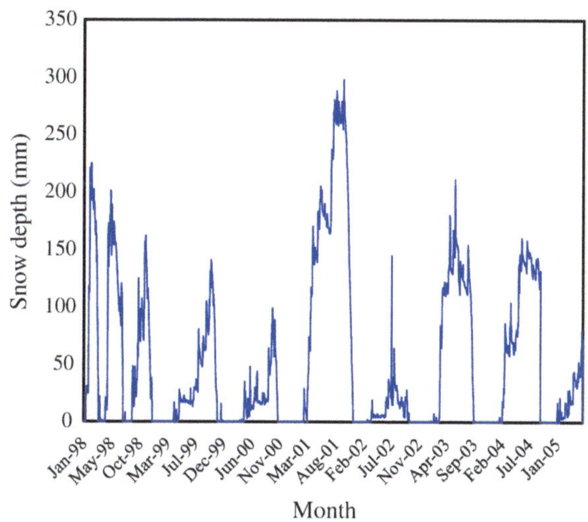

Fig. 4.9 Maximum snow depth between 1995 and 2005

4.5 Hydrology–Discharge

According to an investigation of Weber [6], the Ubaye River can be hydrologically defined as a "Nivo-Pluvial" regime. The results of analysis of maximum discharge between 1961 and 2009 show that the maximum discharge happened in May 2008 with the amount of 205 m^3/s (Fig. 4.10).

Monthly investigation shows that lower discharge occurred during December to March, while the maximum amount of water was recorded from April to July, maybe as a concentrated effect of rainfall and snowmelt together (Fig. 4.11).

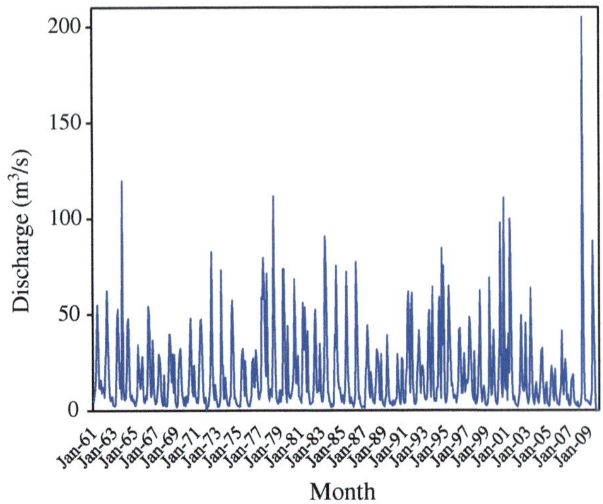

Fig. 4.10 Monthly discharge between 1961 and 2009

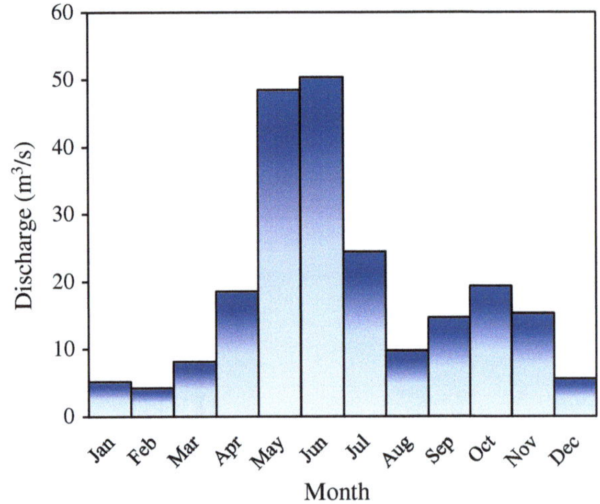

Fig. 4.11 Average monthly discharge between 1961 and 2009

In October, another peak was recorded due to autumn rainfall. The various streams that contribute to the main river are torrential in nature, and seasonal torrentiality of the Ubaye can be attributed to the pattern of the rainfall in the area. This also results in a high rate of erosion and accumulation of debris and sediments downstream. Torrential flood flows occurred generally from April to November, with the peak during May and June when spring floods arise due to excess rainfall and a higher rate of snowmelt.

April, May, June, and July were used to compare the discharge value with precipitation and temperature, and they are sensitive months with regard to the discharge parameter.

By observing the results of May 2008, when the maximum discharge occurred, it can be seen that the precipitation was 156 mm and the maximum temperature

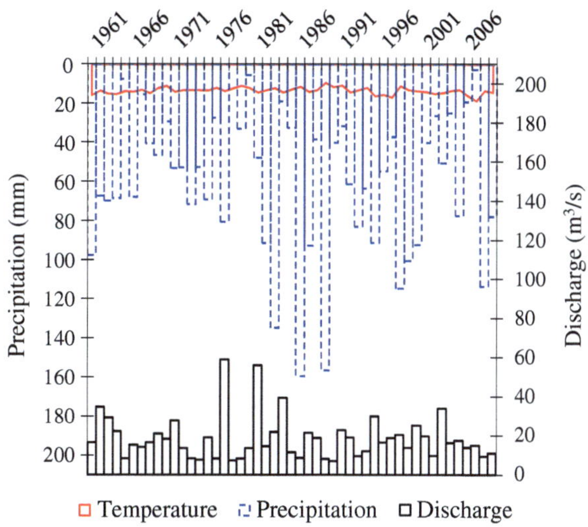

Fig. 4.12 Comparison between precipitation, temperature, and discharge in April (Between 1961 and 2009)

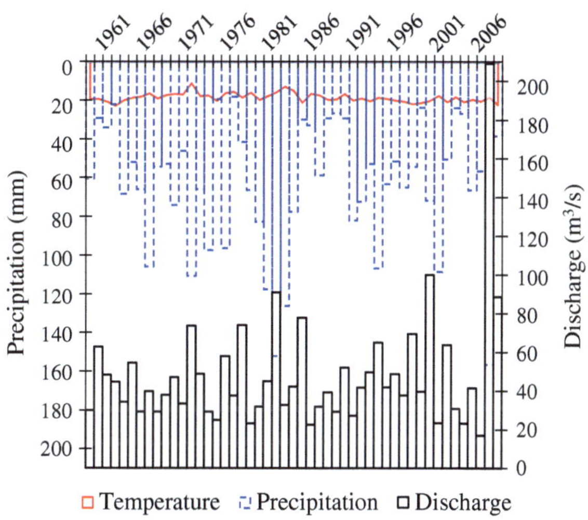

Fig. 4.13 Comparison between precipitation, temperature, and discharge in May (Between 1961 and 2009)

was 22.9 °C. To better compare the three variables plotted as Figs. 4.12, 4.13, 4.14, and 4.15.

The results of time series analysis between 1961 and 2009 show that the highest precipitation was in 1977. Therefore, this year was selected to compare with 2008 (the year of recent flood occurrence). The results of the analysis for monthly data of precipitation, discharge, and temperature are shown in Table 4.2 and Fig. 4.16.

The following results were determined from Fig. 4.16a, b and Table 4.2:

- The precipitation for almost all the months in 1977 was more than 2008. Generally speaking, precipitation decreases at the end of the period (1961–209).
- The temperature for almost all the months in 2008 represented an increasing amount to maximum 4 °C.

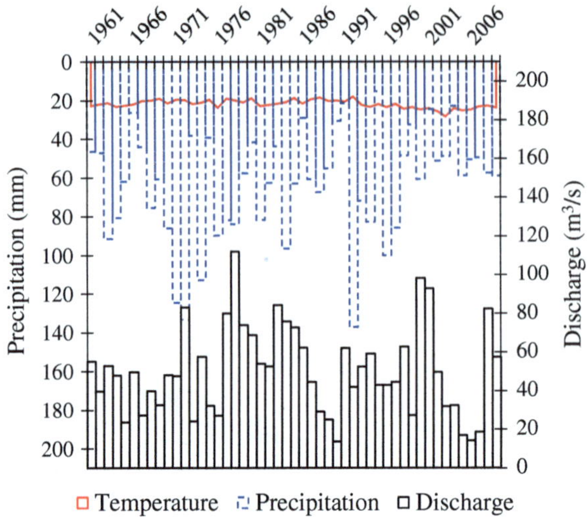

Fig. 4.14 Comparison between precipitation, temperature, and discharge in June (Between 1961 and 2009)

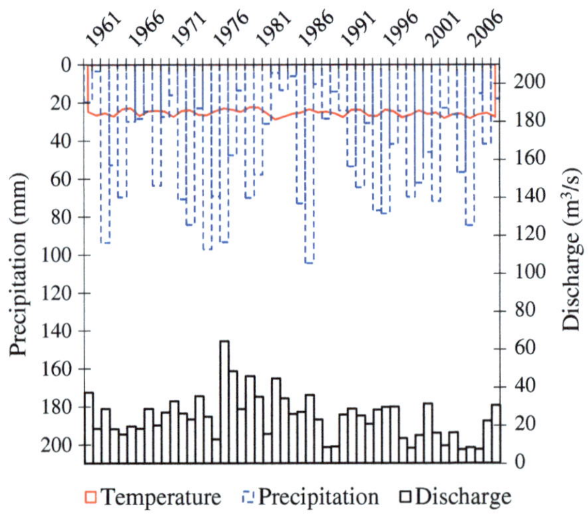

Fig. 4.15 Comparison between precipitation, temperature, and discharge in July (Between 1961 and 2009)

- Maximum discharge in 2008 was more than 1977.

According to these results, the proposed question is "why despite the abundant rainfall in 1977, the amount of discharge was less than 2008?" It is assumed that the land use change and melting of snow as triggering factors are an acceptable reason which is addressed in Chap. 6.

Table 4.2 Comparison between precipitation (mm), Maximum temperature (°C), and discharge (m^3/s) for 1977 and 2008

Month	Parameters	1977	2008
January	Precipitation (mm)	276.9	66.6
	Discharge (m^3/s)	6	1.8
	Temperature (°C)	5.8	7.9
February	Precipitation (mm)	311.5	29.4
	Discharge (m^3/s)	8.45	2.6
	Temperature (°C)	6.06	10.2
March	Precipitation (mm)	346.1	32.6
	Discharge (m^3/s)	11.30	3.1
	Temperature (°C)	6.04	10.3
April	Precipitation (mm)	80.5	113.6
	Discharge (m^3/s)	59.10	9.4
	Temperature (°C)	13.61	13.6
May	Precipitation (mm)	96.1	156.0
	Discharge (m^3/s)	58	205
	Temperature (°C)	16.04	18.1
June	Precipitation (mm)	81.8	57.4
	Discharge (m^3/s)	79.90	82.3
	Temperature (°C)	19.03	22.8
July	Precipitation (mm)	93.4	41.7
	Discharge (m^3/s)	64	22.2
	Temperature (°C)	22.95	25.5
August	Precipitation (mm)	116.1	27.9
	Discharge (m^3/s)	27.60	5.6
	Temperature (°C)	20.93	26.2
September	Precipitation (mm)	18	43.6
	Discharge (m^3/s)	18	4.8
	Temperature (°C)	19.01	20.4
October	Precipitation (mm)	201.2	57.4
	Discharge (m^3/s)	71.7	5
	Temperature (°C)	15.50	17.2
November	Precipitation (mm)	49.8	59.4
	Discharge (m^3/s)	18.30	5
	Temperature (°C)	9.55	8.8
December	Precipitation (mm)	110.2	57.4
	Discharge (m^3/s)	8.06	3.6
	Temperature (°C)	6.38	3.8

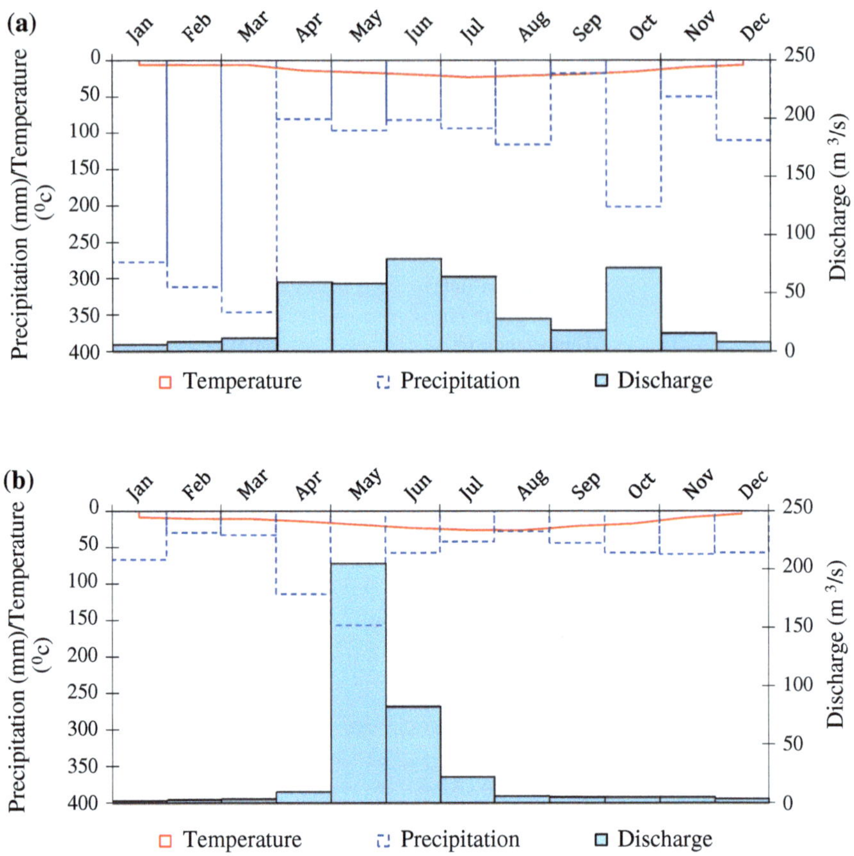

Fig. 4.16 Comparison between precipitation (mm), maximum temperature (°C), and discharge (m^3/s) for 1977 and 2008

4.6 Land Use

In Chap. 2, the land use change process is completely discussed; therefore, a very short description about land use change and the current situation are described in this section. The landscape in the Barcelonnette Basin closely resembles a mass-wasting process and/or badlands. It is obvious that the landscape has primarily been affected by the many geomorphological processes. There is, however, a direct relationship between the landscape development and the geomorphological process. The landscape is characterized by many landslides, torrents, debris fans, and it is also influenced by human activity such as reforestation or the installation of check dams.

According to Weber's investigation (1994), sheep breeding played a very important role in the rural economy of the Barcelonnette Basin for a long time. Every summer, shepherds would take their flocks up to alpine pastures to graze. This resulted in deforestation (15–16th centuries) of high altitude areas for cultivation, which in turn caused considerable damage to the land because of soil erosion, debris flow, and increasingly frequent flooding. To combat these detrimental effects, local laws were put in place to regulate the amount of wood cutting and sheep grazing. During the twentieth century, extensive reforestation took place in the entire Ubaye Valley.

The Restauration des Terrains en Montagne (RTM) started a vast operation of reforestation and construction of dams along the torrents. This development influenced the occurrence and intensity of the hazards and helped reduce them. Barcelonnette is a small town but nevertheless plays an important administrative, commercial, and communication role, providing many tourist activities, hotels, a hospital, and hosts various industries. The mountains are very attractive for skies and hikers. Therefore, the town of Barcelonnette adjusts to these demands, building ski-fields, walking tracks, roads, hotels, etc.

Due to the deforestation of large land areas and the resulting consequences, reforestation started in 1864 based on reforestation laws in the area. These activities included the reforestation of 25,000 ha, the construction of several hundred barriers, and the installation of drains. The benefits of these activities were seen after 1887, when debris flow began to disappear, especially between 1929 and 1948. However, because these constructions were not maintained, they started to deteriorate. As a consequence, there was an increase in flood hazards after 1950, with a centennial flood of the Ubaye River in June 1957 [8].

The continuous changes in the land use in terms of clearing for agricultural practices in the last two centuries and instability of climatic factors have also caused damages in the area [8]. Agricultural activities started in the Roman period and reached their peak in the second half of the last century, so much that the slopes have almost been completely cleared for cultivation resulting in more and more flooding, landslides, and debris flows which lead to catastrophic damages. To reduce them, the Restauration des Terrains en Montagne office (RTM) started a vast operation of reforestation and construction of dams along the torrents.

Now, the current land use category in this area mostly includes coniferous forests (55.94 %), broad-leaved forests (4.85 %), natural grassland (0.8 %), arable land/permanent crops (13.64 %), pastures (8.38 %), bare rocks (1.6 %), black marls (2.86 %), urban fabric (11.24 %), airdrome (0.45 %), water course (0.21 %), marshes, and water bodies (0.01 %) (Fig. 4.17).

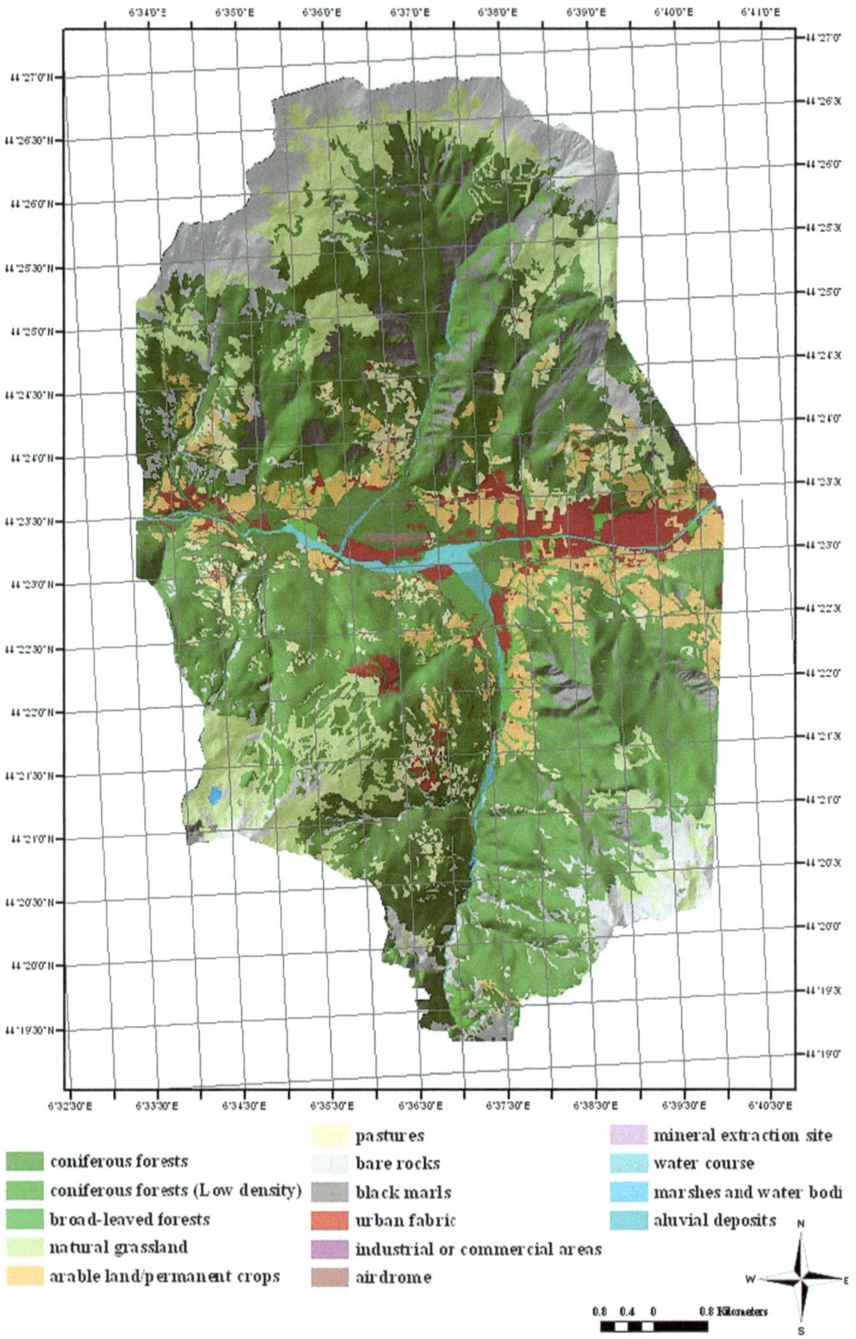

Fig. 4.17 Different categories of land use in Barcelonnette area (Digitized based on available aerial photo)

4.7 Geology and Geomorphology

The Barcelonnette Basin is geologically characterized by two Eocene sheet thrusts (Autapie and Parpaillon), made of strong limestones or sandstones, and overlaying an autochthonous bedrock of soft autochthonous Oxfordian–Callovian black marls. The thickness of the black marls reaches 250–300 m. The black marls have a thickness of 250–300 m and can be divided into three subsets: (a) The Lower Callovian black marl with detrital plates (50–60 m thick); (b) The Middle and Upper Oxfordian black marls (150–250 m thick); (c) with centimeter argillaceous-limestone banks; Some rare outcrops of Argovian black marl (15–20 m thick) [10].

Due to this predisposing geological structure, among some affects of this geological structure are: the hillslopes are affected by (a) severe gullying; (b) shallow landslides; and (c) deep-seated large landslides (Fig. 4.18).

The geomorphological (Fig. 4.19) setting of the Barcelonnette basin is complex, and has strong control of landslide distribution. The complexity of this geomorphological setting has a strong control of landslide distribution [10, 11].

Hilltops consist of very steep slopes (slope gradient ranging between 30° and 70°) and more gentle slopes (slope gradient ranging between 10° and 20°). Alternating

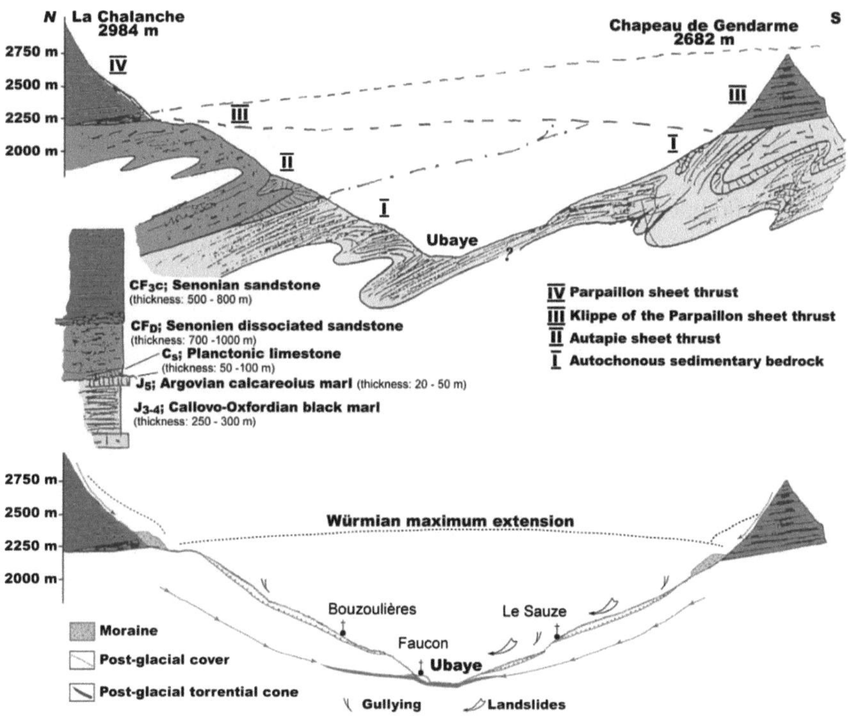

Fig. 4.18 Geological and geomorphological cross-section of the Barcelonnette Basin [10]

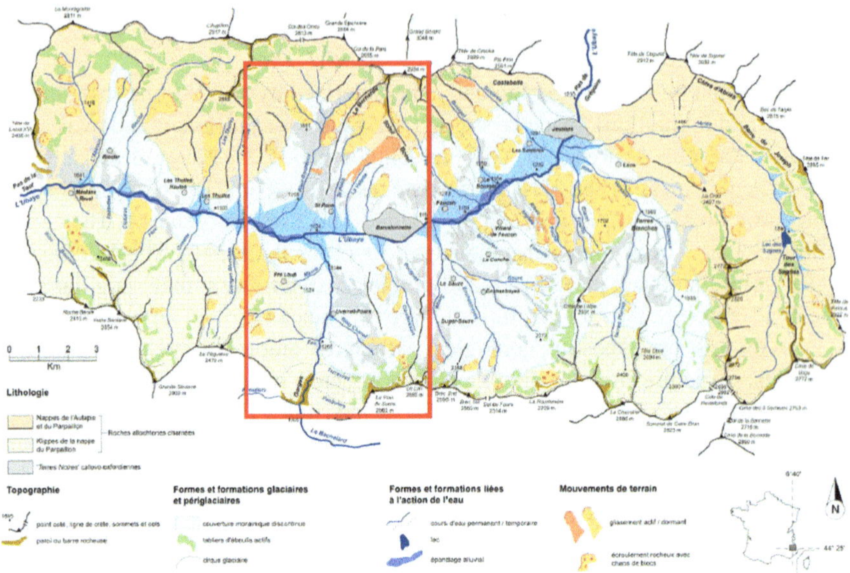

Fig. 4.19 Geomorphology map of Barcelonnette Basin; Inside of red box shows the case study (*source* [10])

steep convex slopes, regular planar slopes, and hummocky slopes all contribute to an irregular relief:

- The steepest convex slopes (>35°) are carved in the black marl outcrops and are very often gullied in badlands or affected by rock-block slides or complex slides.
- Thick morainic deposits (between 10 and 20 m) make up the regular planar slopes, which are often cultivated and affected by rotational and translational slides.
- Forests and/or natural grasslands cover hummocky slopes [10].

Consequently, the main landforms are characterized by (a) large relict landslides in moraine; (b) badlands in black marls; (c) deep-seated mudslides in black marls; and (d) shallow slumps or debris slides in the moraine or in the colluvial cover. Glacial erosion activity has further contributed to carving out the landscape [6, 11, 12].

The varied history of earth movements such as landslides, mudslides, and debris flows in the Barcelonnette area results in features such as "Terres Noires," a predominantly marly dark formation [13]. With a very soft morphology due to the marls, black schist's, and limestone (84 %) at the base of the foot slopes, the area is susceptible to landslides and erosion [6].

Another prominent influence is the high variability of lithologic structures. Erodible structures like black marl and flysch affect the slopes with severe gully erosion and explain the 17 zones of instability. This lithology dates back to the Jurassic period. Mountain areas mostly consist of calcschistes, sandstone, and conglomerate (16 %) with the age start of Cretaceous [8].

Weber [6] explained the presence of Quaternary deposits topped with recent alluvial deposit carpets, saying they begin to form on the surface of dark Limon after being deposited during major floods. It is also important to mention changes in land use from agricultural clearing practices and the unstable climatic factors which have caused damages in the area [8].

4.8 Soil

It is known from past research on marly landscapes that mass-waste processes and rill erosion affect formations with similar morphological and sedimentological characteristics differently. [14–16]. Mass wasting was often found to be prevalent in more cohesive soils because of their higher retention capacity, whereas gully erosion is commonly observed on more silty soils [17]. Extensive erosion from the slopes has contributed to the main types of soils in the area. Weathering produces massive blocks of sandstone and limestone which disintegrates to form sandy and loamy regolith. Flysch produces smaller or massive plates which break down as sandy and loamy regolith, while marls produce silts or clayey regolith in the area [18]. There is presence of some isolated arms of old meanders with clay deposits and moraines in the higher slopes [6].

In order to better understand the soil surface and the influence of infiltration through the surface, several studies have been carried out in the large mudslide area [10]. The case study includes some parts of the Barcelonnette, Faucon-De-Barcelonnette, Saint-Pons, Uvernet-Fours, and Les Thuiles subbasins. Different types of soils (Fig. 4.20) are distinguished in this area, such as moraines, torrential deposits, weathered marls, limestone, and Flysche, which covered 53, 33, 8.13, 3, and 2.8 % of the whole area, respectively. Several factors over the past decades have contributed to torrential floods and debris flows, such as the removal of natural vegetation and increasing tourism. By removing natural vegetation, the likelihood that soil will move down steep slopes increases, resulting in sedimentation of rivers and a higher probability of flash floods.

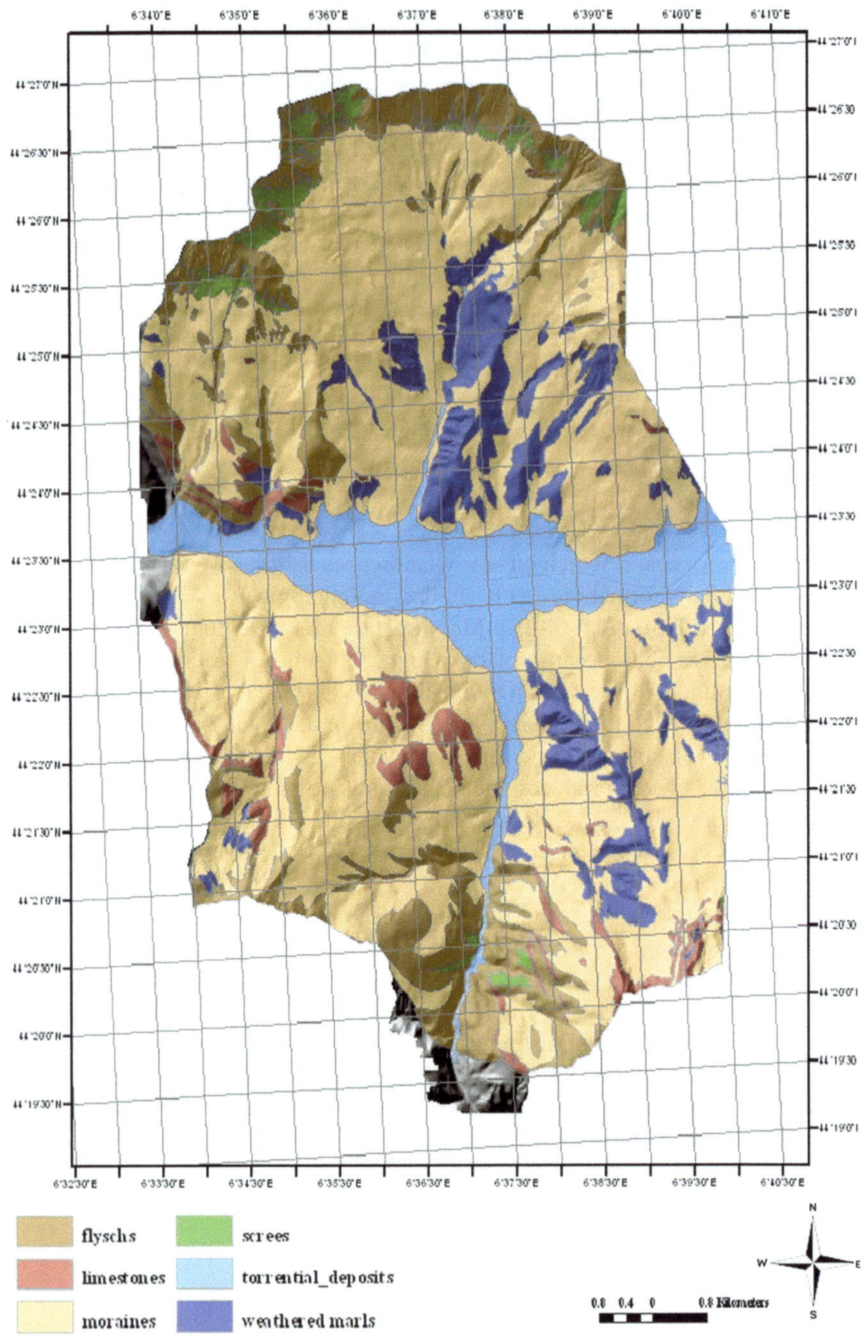

Fig. 4.20 Different types of soil in Barcelonnette area (*source* OMIV)

4.9 Chapter Summary

In this chapter, the main attempt was to introduce the hydro-climatological and environmental characteristics of the study area. The Barcelonnette Basin is affected by a Mediterranean marine influence and moderated by a mild continental mountain climate. During 1961–2009, the cumulative annual precipitation was 1781.6 mm. The analysis of average monthly precipitation between 1961 and 2009 shows that the maximum rainfall occurred in spring and autumn, and the maximum peak happened in May, June, September, October, and November. Based on the observed data at the Barcelonnette station from 1961 to 2009, the maximum temperature (29.37 °C) was recorded in August 2003 and the minimum temperature (−13.59 °C) occurred in January 1981. The maximum temperature occurred between May and September. During the same period, a monthly investigation in discharge shows that the lowest discharge happened from December to March, while the maximum amount of water occurred from April to July, perhaps as a concentrated effect of rainfall and snowmelt together. The results of analysis show that the maximum discharge occurred in May 2008. At that time, the amount of precipitation was 156 mm and the maximum temperature was 22.9 °C.

In general, the Ubaye Valley represents climatic, lithological, geomorphologic, and land use conditions observed in the South French Alps, and is highly affected by flood hazards. The Barcelonnette area is a tourist center with many ski lodges. The whole Barcelonnette Basin is known for its numerous slope (the slope gradients range from 10° to 20°) instabilities and other geomorphological hazards, and the very active flood hazard in this basin directly and indirectly affects land use.

The next chapter provides available data and sources of the data, as well as other information which was applied in this research.

References

1. Flageollet JC, Maquaire O, Martin B, Weber D (1999) Landslides and climatic conditions in the Barcelonnette and Vars basins (Southern French Alps, France). Geomorphology 30:65–78
2. Remaitre A, Malet JP (2010) The effectiveness of torrent check dams to control channel instability: example of debris-flow events in clay shales. In: Garcia CC, Lenzi MA (eds) Check dams, morphological adjustments and erosion control in torrential streams. Nova Science Publishers, New York, pp 211–237
3. Parde M (1925) Le regime du Rhine, etude hydrologique. Institut des Etudes Rhodaniennes, Lyon(In French), pp 883 and 440
4. Piegay H, Salvador P-G (1997) Contemporary floodplain forest evolution along the middle Ubaye River. South Alps Fr Global Ecol Biogeography Lett 6:397–406
5. Piegay H (1995) Dynamiques et gestion de la ripisylve de cinq cours d'eau a charge grossiere du bassin du Rh6ne (I'Ain, I'Ardeche, le Giffre, l'Ouveze et l'Ubaye), XIXeme- XXeme Siecles, 529 pp. Th&se de G&ographie et Amenagement, Universite Paris IV-Sorbonne, Paris (In French)
6. Weber d (1994) Research into earth movements in the barcelonnette basin.In: Casale R, fantechi R, Flageollet JC (eds) Temporal occurrence and forecasting of landslides in the

European Community, Final report, Contract EPOCH, European Commission, vol 1. pp 321–336
7. Henry ST (2010) Flash flood scenario modelling for preparedness and mitigation; case study of Barcelonntte, France, Master Thesis. International Institute for Geo-information Science and Earth Observation(ITC), Twente University, Enschede, The Netherlands
8. Flageollet JC (1996) The time dimension in the study of mass movements. Geomorphology 15:185–190
9. Bhattacharya N (2010) Flood Risk assessment in Barcelonnette, France, M.Sc. Thesis, International Institute for Geo-information Science and Earth Observation(ITC), Twente University, Enschede, The Netherlands
10. Malet, JP, Maquaire O, Remaitre A, Thiery Y, Peyron M, Waeckel C (2010) Landslide hazard and risk assessment in the barcelonnette basin, mountain risks, research training networks, Causeries de L'e'te'—Barcelonnette, 23rd June 2010
11. Remaître A, Maquaire O, Pierre S (2002) Zones d'initiation et de contribution des laves torrentielles dans les bassins marneux. Exemple du torrent de Faucon (Bassin de Barcelonnette, Alpesde- Haute-Provence). Géomorphologie: Relief, Processus, Environ 2002–1:71–84
12. Malet JP, Maquaire O, Calais E (2002) The use of global positioning system for the continuous monitoring of landslides. application to the super-sauze earthflow (Alpes-de-Haute-Provence, France). Geomorphology 43:33–54
13. Antoine P, Giraud A, Meunier M, Vand Asch Th WJ (1995) Geological and geotechnical properties of the "Terres Noires" in southeastern France: Weathering, erosion, solid transport and instability. Eng Geol 40:223–234
14. Gerrits J, Imeson AC, Verstraten JM, Bryan RB (1987) Rill development and badland regolith properties. Catena Suppl 8:141–160
15. Imeson AC, Verstraten JM (1988) Rills on badland slopes: a physico-chemical controlled phenomenon. Catena Suppl 12:139–150
16. Maquaire O, Ritzenthaler A, Fabre D, Thiery Y, Truchet E, Malet J-P, Monnet J (2002) Caractérisation des profils de formations superficielles par pénétrométrie dynamique a' énergie variable: application aux marnes noires de Draix (Alpes-de- Haute-Provence, France). C R Geosci 334:835–841
17. Martínez-Mena M, Castillo V, Albaladejo J (2002) Relations between interril erosion processes and sediment particle size distribution in a semiarid mediterranean area of Southeast Spain. Geomorphology 45:261–275
18. Blijenberg H (1998) Rolling stones? Triggering and frequency of hillslope debris flows in the Barchelard valley. Utrecht University, Southen French Alps

Chapter 5
Available Data

5.1 Introduction

This chapter covers the available data and sources of the data and other information which was applied in this research. All available data such as the digital elevation model (DEM), aerial photos, GIS layers such as land use maps, soil map, hydro maps, geological maps, and hydrometric and climatological data were provided by Dr. Jean-Philippe Malet from the Meteorological Department in Barcelonnette and Observatoire Multidisciplinaire des Instabilités de Versants' (OMIV): [1]. The radar-derived DEM of 5 and 10 m resolutions was retrieved from the OMIV coordinator, OSUG-LGIT, Maison des Géosciences in France.

5.2 Data Description

Aerial Photographs: Except for 1956 and 1995, all aerial photographs from 1974, 1982, 2000, and 2004 (Table 5.1) were orthorectified. The orthorectification process for aerial photos from 1956 to 1995 was carried out as a project study by Rainer Stumper, a Geography student at Vienna University. The sequence of the photographs from 1995 consists of 45 images with a minimum scale of 1:28.000. By a scan resolution of 600 ppi it represents a spatial resolution of at least 1.0 m per pixel. The aerial triangulation was done by using the software "ERDAS Imagine 9.1—Leica Photogrammetrie Suite". The camera parameters were set and the images were initially arranged in their exterior orientation.

The correct arrangement of the flight plan was important for the iteration progress, which will not calculate accurate orientation by wrong settings. To create the geometric relation, 108 tie points and 34 GCPs were identified on the images and the orthophotograph from 2004 respectively. Their adjustment and distribution were chosen after the schema from the ERDAS manual [2]. The result of the iteration was a total RMSE of 1.79 pixels. With this model the images were

A. Ramesh, *Response of Flood Events to Land Use and Climate Change*,
Springer Theses, DOI: 10.1007/978-94-007-5527-7_5,

rectified and resampled individually and later mosaiced with a resolution of 0.75 m per pixel in ArcGIS 9.3.

As a resample method the "cubic convolution" was chosen after visually testing against the others. For aerial Photographs from 1956, the orthorectification process of the images from this year is principally equal to that from 1995. For this year, 29 images were available with a minimum scale of 1:39.400. By a scan resolution of 600 ppi this represents a spatial resolution of at least 1.7 m per pixel. As in the previous section, the parameters were fixed, but only the focal length of the camera is known and can be used. The geometric relation was established by 109 tie points and 40 GCPs which were again identified on the orthophotograph from 2004. The result of the iteration was a total RMSE of 3.16 pixels. The spatial resolution for these photographs was fixated with 1.5 m per pixel [3].

Other available data useful for modeling and analysis hydrometric and meteorological data are explained in detail as follows:

- *Hydrometric data*: this included hourly discharge (m^3/s) and was recorded at the measurement station in Abattoir in Barcelonnette, only on 01.01.1981, 01.01.1982, 01.01.1983, 01.01.1984, 01.1.1985, 01.01.1986, 01.01.1987, 01.10.1988, 30.09.1991, 17.03.1995, 04.05.1999, 25.10.1999, 17.12.2004, 19.05.2006, and 30.05.2008. Monthly discharges were available from 1969 to 2009.
- *Hydrometric data*: this included hourly height of water (m) and also was recorded at the measurement station in Abattoir in Barcelonnette for the period of 1980–2009.
- *The meteorological data*: applied in this research, including precipitation and temperature, provided by the meteorological organization in France, are as follows:

  a. Hourly precipitation (mm) between 1998 and 2009 and monthly precipitation from 1961 to 2009.
  b. Monthly precipitation for Barcelonnette (1961–2009); Jausiers (1961–2009); Uvernet (1954–2009); Condamine (1954–2004); Allos-Nord (1971–2009); SaintPaul (1971–2009).
  c. Daily and monthly temperature data (°c) (Max, Min, Mean) between 1961 and 2009.
  d. Daily and monthly maximum snow depth (1995–2005).

All available raw data are mentioned in Table 5.1. These data were applied for analyzing and modeling in this research.

Additionally, the local risk protection authority—Service de Restauration des Terrains de Montage—Mountain and Restoration Service (RTM) and Plan d'Exposition aux Risques (PER), in combination with the Municipality, had prepared maps for risk assessment based on multiple natural hazards for the case study region. Prévention des Risques Naturels (PPR) also provided the flood hazard zone since 1995 onward. These two maps provided by Dr. Jean-Philippe Malet from RTM, PER, and PPR, were applied in this research to compare the flood simulation results.

Table 5.1 Available data: digital elevation model (file format: Ascii), aerial photos, GIS layers (Shp file and GeoTiff), hydrometric and meteorological data

Name	Definition	Scale
DEM-Barcelo-10 m	Digital elevation model (10 m grid)	1/10.000; 10 m
DEM-Barcelo-5 m	Digital elevation model (5 m grid)	1/5.000; 5 m
Ortho—1956	Non rectified aerial photographs	1/25,000
Ortho—1974	Ortho-photograph coverage	1.0 m
Ortho—1982	Ortho-photograph coverage	1.0 m
Ortho—1995	Non rectified aerial photographs	1/20,000
Ortho—2000	Ortho-photograph coverage	0.5 m
Ortho—2004	Ortho-photograph coverage	0.5 m
Land use maps (1974, 1982, 2000)	Land cover map	1/10.000; 10 m
Soil map	Soil and surficial formation map	1/10.000; 10 m
Hydro map	(Stream network)	1/10.000; 10 m
Geological map (1974)	Detailed geological formation map	1/50.000; 50 m
Height of water (m)	Time series at Barcelonnette	Hourly, monthly
Discharge (m^3/s)	Time series at Barcelonnette	Hourly
Discharge (m^3/s)	Time series at Barcelonnette	Monthly
Precipitation (mm)	Time series at Barcelonnette	Hourly
Precipitation (mm)	Time series at Barcelonnette	Monthly
Precipitation (mm)	Time series at Jausiers	Monthly
Precipitation (mm)	Time series at Uvernet	Monthly
Precipitation (mm)	Time series at condamine	Monthly
Precipitation (mm)	Time series at Allos-Nord	Monthly
Precipitation (mm)	Time series at Saint Paul	Monthly
Temperature (max, min, mean) (°c)	Time series at Barcelonnette	Monthly
Snow depth (mm)	Restofond	Hourly, monthly

5.3 Chapter Summary

The radar-derived DEM of 5 and 10 m resolutions were retrieved from the OMIV coordinator, OSUG-LGIT, Maison des Géosciences in France. Except for 1956 and 1995, all aerial photographs in 1974, 1982, 2000, and 2004 were provided by French organizations. The orthorectification process for aerial-photos from 1956 and 1995 was conducted at Vienna University. All other available data useful for modeling and analyses are included: hourly discharge (m^3/s) (limited only to 14 measurements between 1981 and 2008); hourly heights of water (m) for the period of 1980–2009; hourly precipitation between 1998 and 2009; monthly precipitation, discharge and temperature from 1961 to 2009. These data were provided by the Met. Dept. Barcelonnette and OMIV.

The next chapter explores all the achieved results of SOBEK and HEC-RAS simulation, climate change detection, and statistical analysis for land use change.

References

1. The French Observatory—Observatoire Multidisciplinaire des Instabilités de Versants (OMIV), Data Access—Barcelonnette area, http://eost.u-strasbg.fr/omiv/data_access_Barcelonnette
2. ERDAS User's manual (2005) Leica geosystems geospatial imaging (LLC), 5051 Peachtree Corners Circle, Suite 100, Norcross, GA, 30092, USA
3. Natural hazards and risk in the barcelonnette area, project report (2009) Southern France, department of geography and regional research, University of Vienna

Chapter 6
Results

6.1 Introduction

Based on the objectives of the dissertation, the results of interpolation methods will first be discussed. The main aim here is to find interpolation methods which give the best results based on statistical methods such as the root mean square (RMSE) and watershed and stream network delineation. Then, two integrated hydrodynamic models (SOBEK and HEC-RAS) will be applied to investigate changes in the flood hazard and to compare different flow characteristics over time in the Barcelonnette basin, with the aims of: first, to model different hydrodynamic scenarios for civil protection purposes; second, to simulate probability distribution and the return period analysis and third, to determine and compare different hydraulic characteristics (i.e., propagation rate, flow depth, and inundation time).

In the subsequent section, the results of statistical methods will be presented in order to detect trend and to find probable abrupt points in spatial distribution of the observed time series (climatological and hydrological parameters). Additionally, to detect land use changes from the past until the present and to ensure any affects on the flow regime of the Ubaye River, the results of the GIS-based process analysis will be shown. The majority of flood events depicted in this research are associated with the latter cause of flooding. Therefore, the last and main objective is to ensure a scientific relationship between climatic and hydrologic parameters, land use and flood events over time. In this regard, the integration and estimation of all achieved results seem to be necessary.

A. Ramesh, *Response of Flood Events to Land Use and Climate Change*,
Springer Theses, DOI: 10.1007/978-94-007-5527-7_6,

6.2 DEM Quality Control Modeling

6.2.1 Interpolation Techniques

In this section, eight digital elevation models (DEMs) were generated using different interpolation techniques and different parameter settings. The spatial resolution of all surfaces was set to 5 m. Three DEMs were generated based on deterministic techniques including IDW, Spline, and Topo to Raster (using the ANUDEM program developed by Michael Hutchinson), and five DEMs were generated using ordinary Kriging methods including Circular, Exponential, Gaussian, Stable, and Spherical.

To select the best DEMs generated with different interpolation techniques, a set of critical measurements were used to compare the results. In the case of the Geostatistic model—Kriging-based method—the selection of an optimal set of parameters to be used in ordinary Kriging required an exploration of different combinations of the values. Different combinations of the lag size and the number of lags were used in the generation of empirical semi-variograms for different surfaces, interpolated as well. Even though the selection of an appropriate lag size depends on the particular phenomenon being modeled, too large of lag sizes may mask short distance autocorrelation, and too short of lags can create in the predictions. In this study a variety of lag sizes were tested, and finally the best results were obtained using a relatively small lag size of 100 and 25 m. One aspect of the interpolation methods that was qualitatively quantified was time efficiency. For the training dataset used in this study, the surface interpolation, with a final resolution of 5 m using Kriging took an average of 1 h for each test. IDW, Topo to Raster and Spline methods were the fastest interpolation techniques (less than 10 min) to interpolate the surfaces to the final resolution. The defined lag size suggests that given the complex topographic conditions of the study area, the autocorrelation between locations is significant only in reduced neighborhoods beyond which conditions change abruptly. Some important conditions were maintained constant in all deterministic and geostatistic models. First, no anisotropy was considered in any of the models. Second, four-sector circular neighborhoods were used in all models and four points per sector were used to make predicted points at unknown locations. Therefore, in this section, two assumptions are considered and the related results are discussed below:

As for the first assumption, interpolation methods that often assume data points in original DEM are correct (exact), but it can be assumed that they are subject to error (generally of a known or estimated extent). This means the interpolation technique is not only for creating DEM from contour lines or observed data points, but also for when we have a DEM that we could correct or find error with, such as sink errors. Therefore, the main purpose was to correct the original DEM and find errors and to address the best interpolation techniques in order to erase errors in DEM. Therefore, 70 % of the observed data points

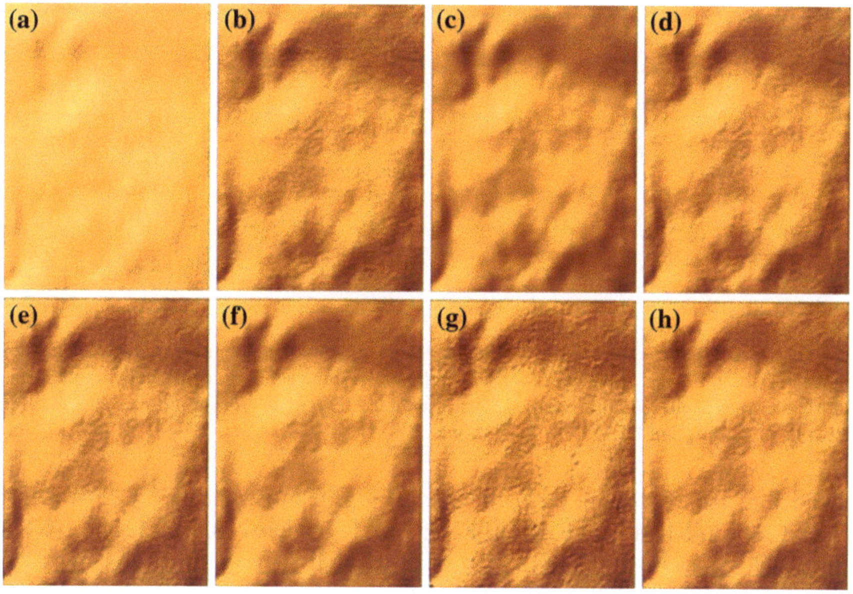

Fig. 6.1 Sample area; comparison between the modeled DEM applied different interpolation techniques. **a** IDW, **b** Topo to Raster (ANUDEM), **c** Spline, **d** Kriging, Circular, **e** Kriging, Exponential, **f** Kriging, Gaussian, **g** Kriging, Stable, **h** Kriging, Spherical

were selected randomly. Then, a DEM was created using different interpolation techniques based on this 70 % data point set.

The DEMs hill-shade images (Fig. 6.1) show a comparison of the final results between different applied interpolation techniques and it focus on a small part of the area.

Statistical analysis: Based on the first assumption, it was assumed that the source DEM has some errors. Due to this reason, there was no observed data to compare the interpolated results statistically.

Hydrological analysis: To compare the accuracy of the original DEM and Modeled DEMs, the stream network for each interpolated surface and original DEM was derived using ArcHYDRO tools in ArcGIS. The modeled stream network was compared with the observed stream network ("true" data). By visualization analysis of all achieved results, it seems that IDW, Spline, Circular, Exponential, and Stable produced the poorest results mainly in the upstream, where a small part of the main channel is missing.

The results obtained from Topo to Raster, Gaussian, and spherical derived drainage networks coincided better with the reference stream network. Still, there is some discrepancy in the modeled network in some parts and the river follows a wrong path. On the other hand, the drainage networks obtained from the above-mentioned DEMs allow a much more accurate delineation of the stream network.

As shown in the Fig. 6.2, the agreement between drainage networks based on IDW and the observed drainage network is not perfect. In particular, a small part

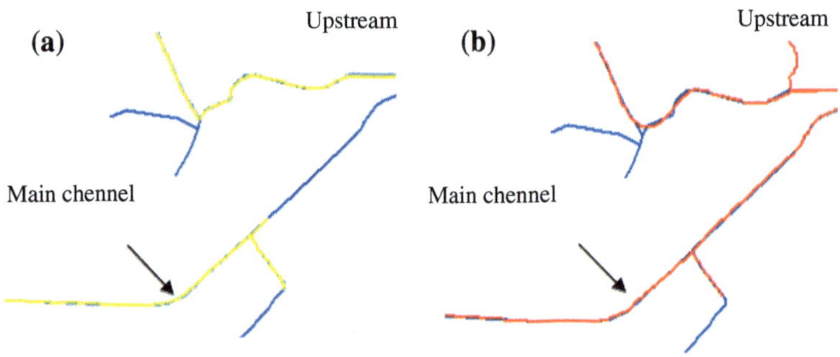

Fig. 6.2 **a** and **b** the agreement between modeled and observed drainage network; *blue line* represents the observed drainage networks; *yellow line* represents drainage network based on DEM created by interpolation methods (as a sample: IDW); *red line* represents the drainage network based original DEM (5 m resolution)

of the drainage network in the main channel (upstream) is missing in IDW DEM. Then again, this part of the drainage network is obviously shown in the original DEM and was more consistent with the observed drainage network. As an overall result, it is concluded that the interpolation techniques were not successful in creating a perfect DEM even more accurate than the original DEM. Consequently, the first assumption is rejected.

Therefore, in the hydrodynamic simulation the original DEM was applied. The quality of DEM was also tested hydrologically using the drainage enforcement algorithm to recognize that each spurious sink is surrounded by a drainage divide containing at least one saddle point. If the sink is associated with an elevation data point, then the lowest such saddle, provided it is not also associated with an elevation data point, identifies the region of the grid, which is modified in order to remove the spurious sink. Figure 6.2 represents the comparison between modeled and observed drainage networks, for IDW compared with the original DEM.

In this part, the second assumption was that the original DEM is included in observed elevation point data with 100 % accuracy. Therefore, it was used as a correct source to test the different interpolation methods. The idea behind this assumption was to find the best interpolation technique which represents the best agreement with the observed elevation of the data point set. Under this premise, the interpolation process which was applied to the first assumption was repeated. Finally, for testing the best method, the elevations for the left-out 30 % data point set, which were created as unknown data points by using interpolation techniques, were compared with the same observed data to see how the accuracy of the DEM decreases as the spatial coverage of high data is reduced. Figure 6.3 represents the process of creating DEM based on different interpolation techniques.

The production of large elevation error estimates, evenly distributed over the DEM, makes it possible to explore some general questions, such as how to identify the statistical characteristics of DEM error, and how much this differs among interpolation methods and the relationship between RMSE of elevation and RMSE

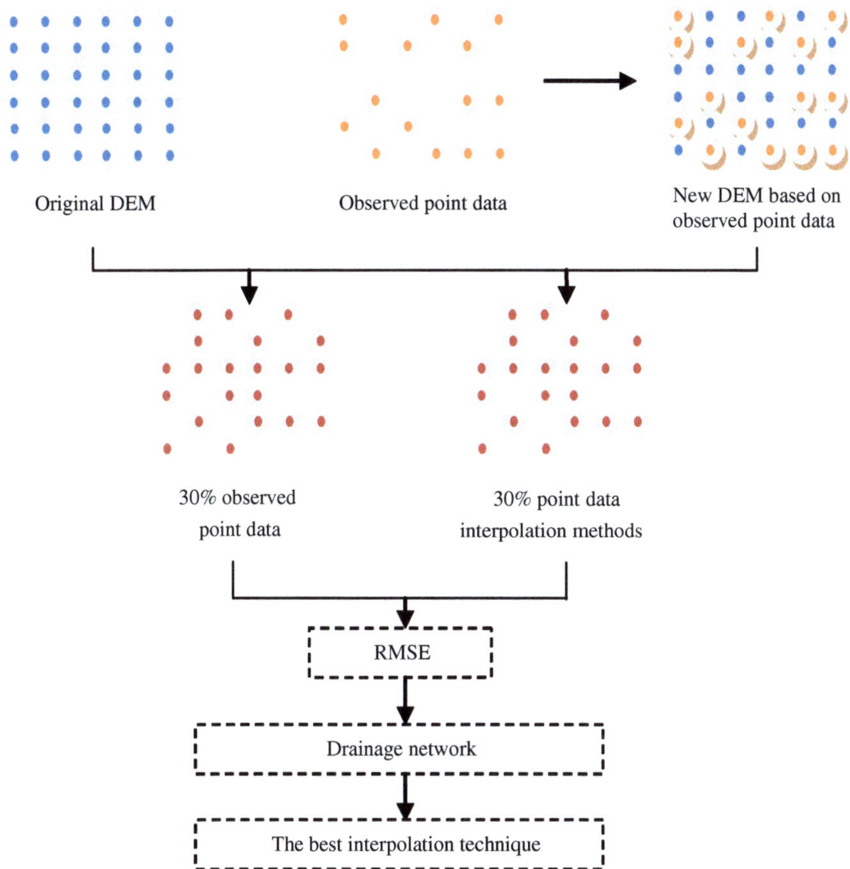

Fig. 6.3 The processing of creating DEM based on different interpolation techniques

of surface derivatives. It should also be pointed out that no consideration is given to how accurate the original DEM values are. When the interpolated values are compared with the original data points, this provides a measure of error introduced by the interpolation process, but says nothing about the absolute accuracy of the elevation values compared with the true ground value.

6.2.2 Statistical Analysis

To quantify the results, first the RMSE was applied. The selection of an optimal set of parameters to be used in deterministic and geostatistical methods required the exploration of different combinations of values. Table 6.1 shows four statistical analyses for each method:

Table 6.1 Accuracy values for the DEMs generated in the geostatistical analyst extension

	Deterministic			Geostatistic				
	IDW	Topo	Spline	Circular	Exponential	Gaussian	Stable	Spherical
RMSE	17.55	17.62	17.54	1.51	.119	.005	.432	.000
Mean	4.18	4.08	4.16	2.98	.092	.003	.230	.000
Max	35.84	36.04	35.85	9.08	.036	.003	.627	.000
R square	.982	.982	.982	1.000	1.000	1.000	1.000	1.000

1. RMSE is calculated for selected data points (randomly 1,000 points)
2. Mean error presents the arithmetic mean of the error values and reveals whether the interpolation has a tendency to under or overestimate on average.
3. Maximum absolute error shows how large the errors can be in extreme cases
4. *R*-square is a statistical measure of how well a regression line approximates real data points; an *R*-squared of "1.0" (100 %) indicates a perfect fit.

As a general conclusion, the results of statistical analysis showed that all DEMs derived from different interpolation techniques were statistically significant. Going into the details in terms of RMSE, better results in model accuracy were obtained when Exponential, Gaussian, Stable, and Spherical models were used. On the other hand, in Geostatistic analysis, the difference was slight and has no bearing on the overall pattern of results. The mean errors in these methods were also close to zero. This means that mean error is simply the arithmetic mean of the error values and reveals whether the interpolation has a tendency to under or overestimate on average. Maximum absolute error also shows that the errors are not so large. *R*-square in all methods shows that the regression line is "1" or close to "1", meaning that in terms of *R*-squared, the results obey a perfect fit especially in Geostatistic methods. In principle, the spatial pattern of errors was similar in all cases of Geostatistic methods. Inversely, in the results of deterministic methods, especially in terms of RMSE, the errors are larger than Geostatistic methods. Therefore, as an overall result, it could be concluded that Geostatistic techniques show the strongest results compared to the deterministic. In between, the Spherical and Gaussian showed better agreement with the observed data and represents the smoothest and more accurate DEM. In terms of visualization pattern of errors underestimation of elevation occurs where the slope is upwardly convex and where a change in elevation is marked (Fig. 6.4). Hence, the largest errors occur along narrow crest lines and at the convex change of slope at the tops of ridges. Large convex areas, such as the tops of broad ridges, are estimated more successfully. Conversely, overestimation occurs in areas that are upwardly concave.

Table 6.2 shows the frequency distribution of error values and Table 6.3 represents the standard error and *P* value of a significant level.

Tests for normality were run on all the distributions and a Kolmogorov–Smirnov test showed that all considered DEMs are significantly different from the original DEM, with a 0.008 *P* value for the Spline and zero for the rest of the applied methods. The standard coefficient and standard error in Geostatistical

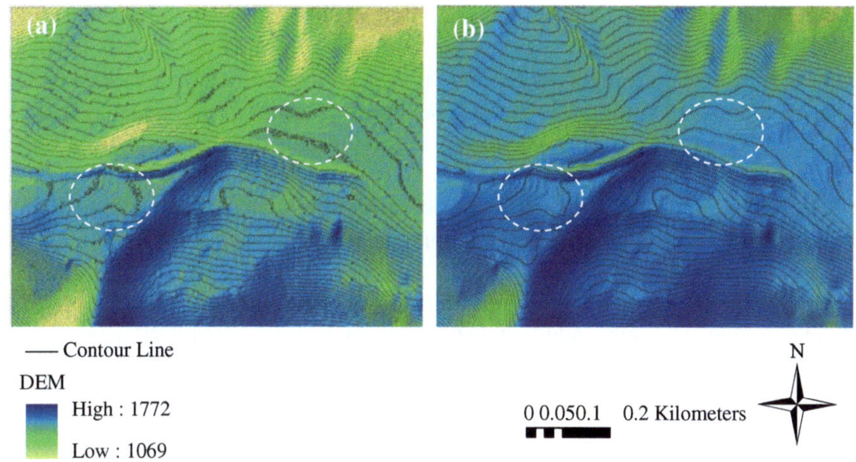

Fig. 6.4 Typical spatial pattern of error. **a** Modeled DEM. **b** Original DEM

Table 6.2 Frequency distribution of DEM error

	Deterministic			Geostatistic				
	IDW	Topo	Spline	Circular	Exponential	Gaussian	Stable	Spherical
Mean	−3.55	−3.55	−3.55	−.000	−.003	−.000	−.024	−.000
Max	2.281	2.303	2.27	.000	.003	.000	.055	.000
R square	.972	.972	.972	1.000	1.000	1.000	1.000	1.000

Table 6.3 Standard error and *P* value of significant level

Variable	Standard coefficient	Standard error	*P* value
IDW	0.986	0.005	0.000
Topo	0.986	0.005	0.000
Spline	0.986	0.005	0.008
Circular	1	<.001	0.000
Exponential	1	<.001	0.000
Gaussian	1	<.001	0.000
Stable	1	<.001	0.000
Spherical	1	<.001	0.000

methods represent better results comparing to deterministic methods. The standard error for all Circular, Exponential, Gaussian, Stable, and Spherical is <.001.

For Geostatistic methods, the standard error was about 0, which means that the spatial distribution of points in applied methods is not too far from the distribution of original values. When modeling elevation error, it is always assumed that error is spatially autocorrelated.

This is a reasonable assumption given that the terrain tends to vary smoothly rather than abruptly, so that the neighboring pixels are likely to have similar height values, and given the fact that interpolation itself is likely to produce similar values for neighboring points. When the number of original data points is small, there will be a very strong autocorrelation between the estimated elevations values since neighboring values are essentially being estimated from the same data points.

The Q–Q plot (Fig. 6.5) quantifies and represents the estimated variable's distribution against the original values more consistent. It could be concluded that in Geostatistic methods the distribution of a variable matches very strongly with the observed or original data distribution. On the other hand, the points cluster around a straight line.

To test the accuracy of different interpolation techniques and to get more idea about the achieved results, additional statistical analysis was estimated using spatial analysis in macroecology (SAM). First, the predicted values for each method were compared with original values of source data points. The minimum, maximum, mean, and standard deviations for each method were calculated and the descriptive statistical results are shown in Table 6.4. These parameters are defined as following:

- Minimum absolute error shows how small the errors can be in case of considering lowest elevation in both observed and estimated value.
- Maximum absolute error shows how large the errors can be in case of considering highest elevation in both observed and estimated value.
- Mean error presents the arithmetic mean of the error values and reveals whether the interpolation has a tendency to under or overestimate on average.
- Standard deviation shows how much variation or "dispersion" there is from the average. A low standard deviation indicates that the data points tend to be very close to the mean, whereas high standard deviation indicates that the data points are spread out over a large range of values.
- "Residuals" are defined as meaning the spatially modeled component of variation not accounted for by the environmental variables, and that is why it has a very strong spatial structure (because it is added when modeling the semi-variogram). By doing this, it is "released", at the same time, the effect of the spatial structure of data points.

In this test the best results also belong to Geostatistic–Spherical method, which achieved the most accurate results with a value of $<.001$. A fundamental task in many statistical analyses is to characterize the location and variability of a dataset. A further characterization of the data includes skewness and kurtosis. To this propose, the histogram of the residuals (Fig. 6.6), including the standard residual diagnosis graphs (Fig. 6.7) was plotted for selected interpolated values. The skewness and kurtosis value for all methods are represented in Table 6.5.

It seems that the dataset for Circular, Exponential, and Gaussian is symmetric, because the distribution is almost the same to the left and right of the center and the skewness in these datasets is close to zero.

The negative values belong to Exponential with the value of -1.69. These negative values indicate data that are skewed left. Skewed left means that the left tail is long relative to the right tail. The positive values belong to Circular and Gaussian. These positive values indicate data that are skewed right. Similarly,

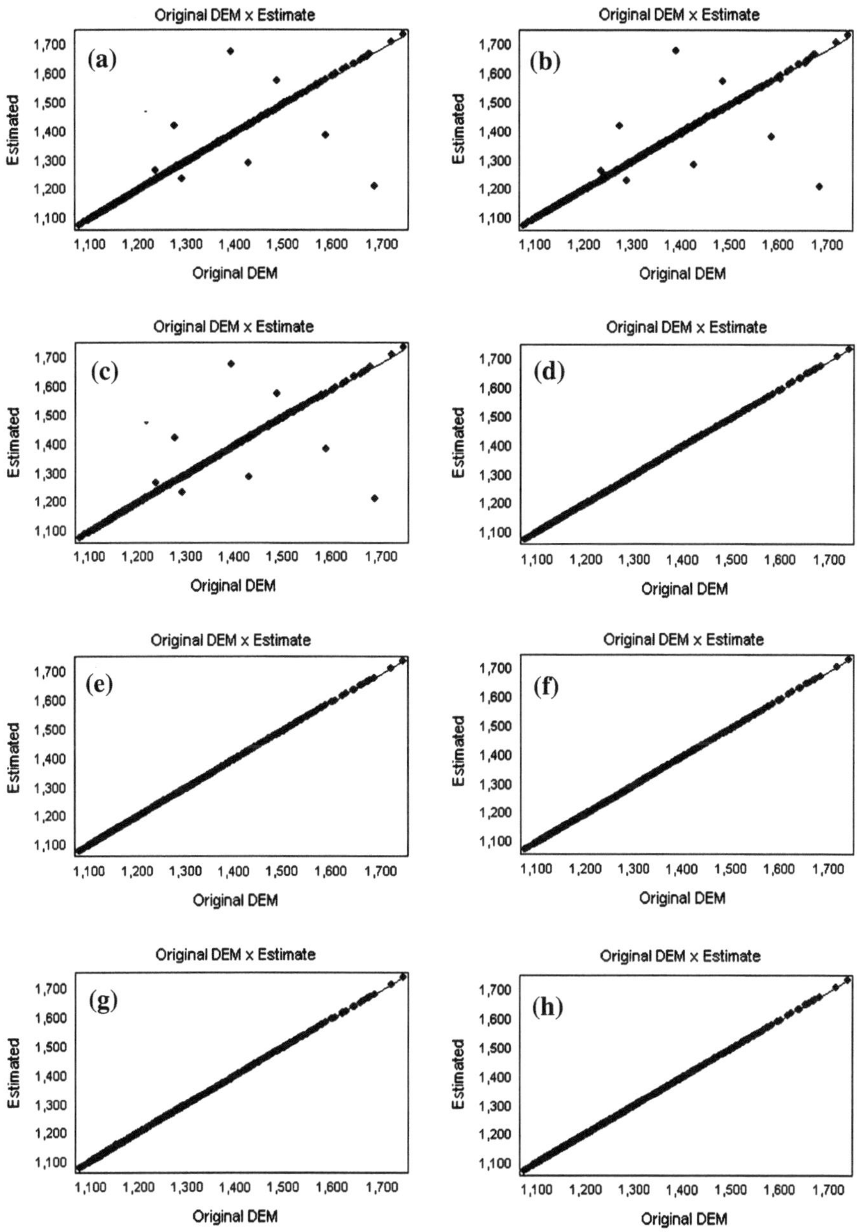

Fig. 6.5 Q–Q plot for error produced using **a** IDW; **b** Topo to Raster; **c** Spline; **d** Circular; **e** Exponential; **f** Gaussian; **g** Stable; **h** Spherical for the first assumption

Table 6.4 Statistical comparison between observed, estimated and residuals values

IDW	Observed	Estimated	Residuals
Min	1073.427	1075.712	−290.332
Max	1739.889	1732.948	470.273
Mean	1246.647	1246.647	<.001
S.D.	130.753	128.928	21.77
Topo to Raster	Original	Estimated	Residuals
Min	1073.427	1075.541	−293.043
Max	1739.889	1735.349	470.222
Mean	1246.647	1246.647	<.001
S.D.	130.753	128.921	21.813
Spline	Original	Estimated	Residuals
Min	1073.427	1075.726	−289.114
Max	1739.889	1735.186	470.319
Mean	1246.647	1246.647	<.001
S.D.	130.753	128.93	21.756
Circular	Original	Estimated	Residuals
Min	1073.427	1073.428	−0.003
Max	1739.889	1739.888	0.004
Mean	1246.647	1246.647	<.001
S.D.	130.753	130.753	<.001
Exponential	Original	Estimated	Residuals
Min	1073.427	1073.424	−0.045
Max	1739.889	1739.878	0.021
Mean	1246.647	1246.647	<.001
S.D.	130.753	130.753	0.005
Gaussian	Original	Estimated	Residuals
Min	1073.427	1073.43	−0.454
Max	1739.889	1739.787	0.452
Mean	1246.647	1246.647	<.001
S.D.	130.753	130.753	0.119
Stable	Original	Estimated	Residuals
Min	1073.427	1073.36	−7.228
Max	1739.889	1740.314	3.177
Mean	1246.647	1246.647	<.001
S.D.	130.753	130.752	0.431
Spherical	Original	Estimated	Residuals
Min	1073.427	1073.427	<.001
Max	1739.889	1739.889	<.001
Mean	1246.647	1246.647	<.001
S.D.	130.753	130.753	<.001

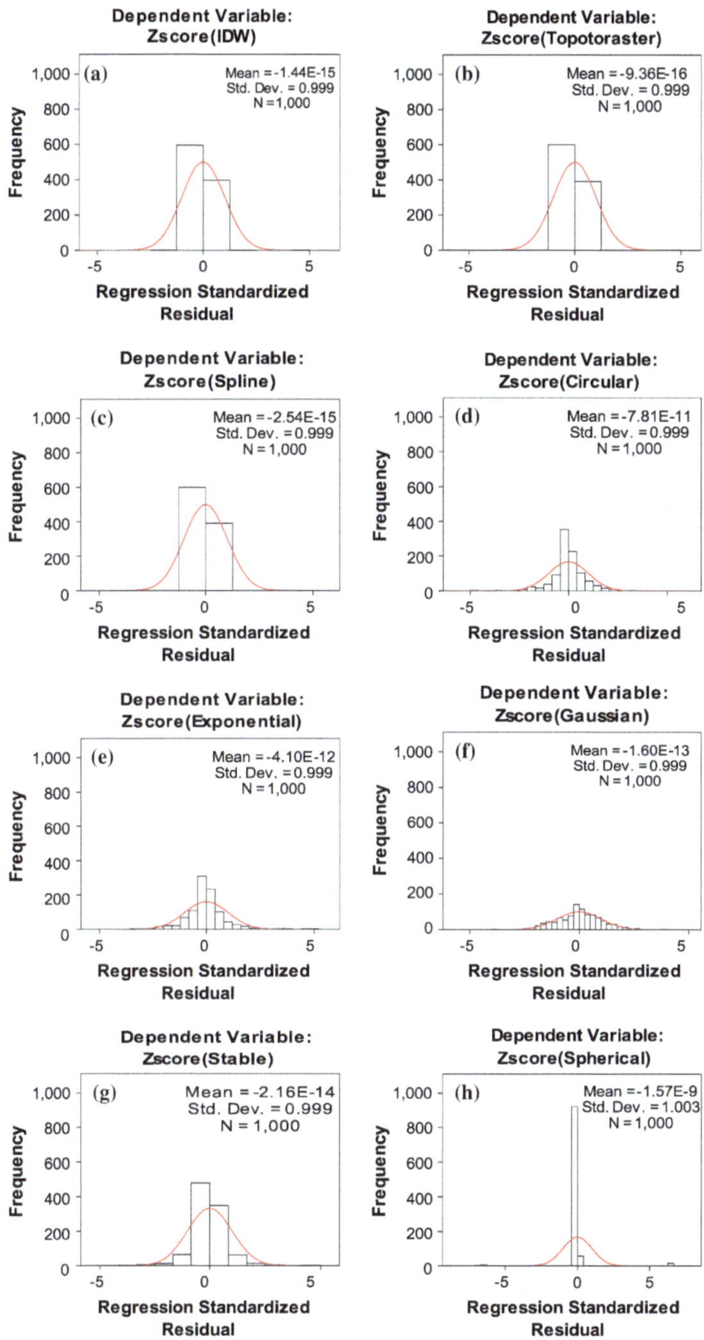

Fig. 6.6 Histogram of the residuals. **a** IDW; **b** Topo to Raster; **c** Spline; **d** Circular; **e** Exponential; **f** Gaussian; **g** Stable; **h** Spherical

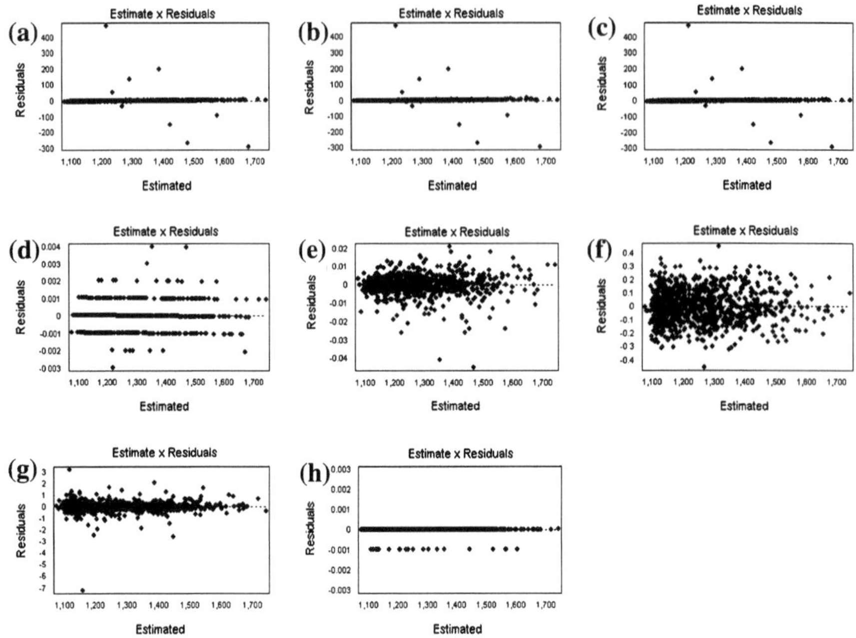

Fig. 6.7 Residual diagnosis graph. **a** IDW; **b** Topo to Raster; **c** Spline; **d** Circular; **e** Exponential; **f** Gaussian; **g** Stable; **h** Spherical

Table 6.5 Skewness and Kurtosis values for all interpolation techniques

Variable	Skewness	Kurtosis
IDW	6.525	261.72
Topo	6.402	280.688
Spline	6.563	282.067
Circular	0.599	4.941
Exponential	−1.69	12.75
Gaussian	0.071	0.193
Stable	−4.702	85.183
Spherical	−6.367	38.645

skewed right means that the right tail is long relative to the left tail. The Standard residual diagnosis graphs show that the best results in terms of normal distribution belong to Gaussian method, because all scattered values spread close to zero (Skewness is 0.071) between the ranges of −0.5 and 0.5 (Fig. 6.7).

Kurtosis is a measure of whether the data are peaked or flat relative to a normal distribution. The dataset for Circular and Exponential with high kurtosis of 4.941 and 12.75 tend to have a distinct peak near the mean, decline rather rapidly, and have heavy tails. Datasets with low kurtosis belong to Gaussian with the value of 0.193 tend to have a flat top near the mean rather than a sharp peak.

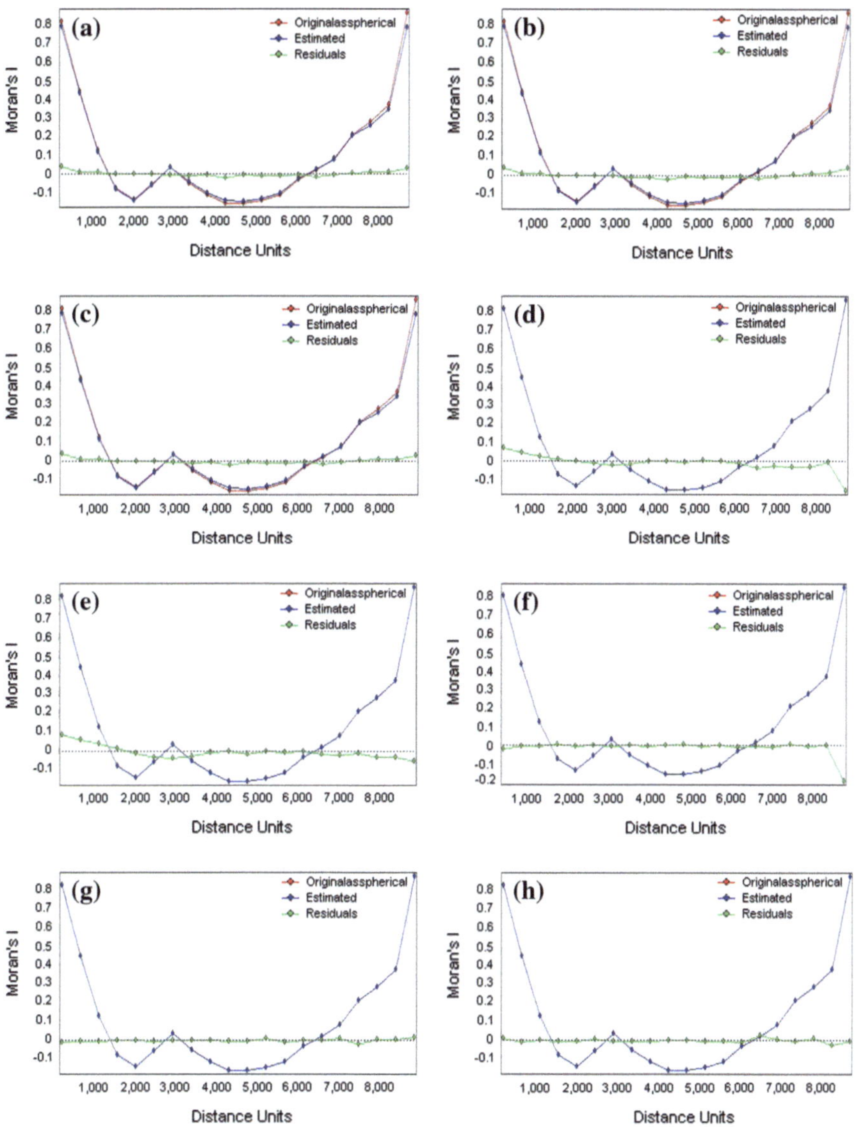

Fig. 6.8 Statistical comparison between original, estimated, and residuals values. **a** IDW; **b** Topo to Raster; **c** Spline; **d** Circular; **e** Exponential; **f** Gaussian; **g** Stable; **h** Spherical

Perhaps the most interesting screens (in terms of what SAM does compare with most software) are exploring spatial patterns in the model, especially in model residuals (the green line of the correlogram, with the red representing the source values and blue representing estimated values) (Fig. 6.8). The results are shown in Table 6.6.

Table 6.6 The high Moran's *I* in the first distance

Method	Moran's *I* value	Result
IDW	0.041	Perfect correlation
Topo to Raster	0.042	Perfect correlation
Spline	0.041	Perfect correlation
Circular	0.071	Perfect correlation
Exponential	0.089	Perfect correlation
Gaussian	−0.019	Perfect dispersion
Stable	−0.008	Perfect dispersion
Spherical	0.013	Perfect correlation

Based on the obtained results, the negative (positive) values indicate negative (positive) spatial autocorrelation. Values range from −1 (indicating perfect dispersion) to +1 (perfect correlation). A zero value indicates a random spatial pattern. The high Moran's I in Gaussian and Stable showing perfect dispersion in the first distance. On the other hand, the Moran's I value for IDW, Topo to Raster, Spline, Circular, Exponential, and Spherical represents perfect correlation in the first distance (Table 6.6).

As partly conclusion from this part, it could be concluded that Geostatistic techniques show the strongest results compared to the deterministic. In between, the Spherical and Gaussian showed better agreement with the observed data and represents the smoothest and more accurate DEM. In terms of spatial analysis, The Standard residual diagnosis graphs show that the best results in terms of normal distribution belong to Gaussian method, because all scattered values spread close to zero (Skewness is 0.071) between the ranges of −0.5 and 0.5. The residuals dataset in Gaussian method obeys a normal distribution. The results of Moran's I investigation showed Gaussian and Stable showing perfect dispersion in the first distance. On the other hand, the Moran's I value for IDW, Topo to Raster, Spline, Circular, Exponential, and Spherical represents perfect correlation in the first distance. In between, the lowest Moran's I value of 0.013 belongs to Spherical.

6.2.3 Hydrological Analysis

To compare the accuracy of the interpolated DEMs, the stream network for each surface was derived using ArcGIS. The modeled stream network was compared with the observed stream network ("true" data). Figure 6.9 shows the results of stream network delineation using ArcHYDRO tools for Kriging–the Spherical model as a sample. In this part, the quality of all the DEM was tested hydrologically by creating drainage networks. Figure 6.9 presents the comparison between the modeled and the observed stream network for (a) original DEM; (b) IDW; (c) Topo to Raster (ANUDEM); (d) Spline; (e) Kriging, Circular; (f) Kriging; Exponential; (g) Kriging, Gaussian; (h) Kriging, Stable and (i) Kriging, Spherical.

Fig. 6.9 Stream network delineation using ArcHYDRO tools

The black line represents the observed stream network and the red line shows the modeled stream network. By reviewing the Fig. 6.9 it seems that overlay of (b) IDW, (d) Spline, (e) Kriging, Circular, (f) Kriging, Exponential, and (h) Kriging, Stable produced the weakest results, mainly in the upstream where a small part of the channel was missing. The results obtained from (c) Topo to Raster, (g) Kriging, Gaussian and (i) Kriging, Spherical is considered to be in better agreement with the observed drainage network. By focusing on details and the interpretation of results, it is concluded that the drainage networks generated by the Kriging, Spherical allowed a much more accurate delineation of the stream network.

As shown in Fig. 6.10, the agreement between drainage networks based on IDW DEM with the observed drainage network is not perfect. In particular, a small part of the drainage network in the main channel (upstream) is missing in the IDW DEM. On the other hand, this part of the drainage network is obviously clear in the Spherical and agreed better with the observed drainage network. Therefore, as would be expected, the Spherical was the best overall modeled drainage network, and it was comparatively more accurate than the river from the DEMs generated with other interpolators, even in the floodplain which is a flat area.

The results of modeled and observed drainage networks for all interpolation methods were estimated visually and tallied into an error matrix (Table 6.7).

Figure 6.11 represents the comparison between modeled and observed drainage networks for the Spherical method.

(a) Representing river in observed not defined as such in the modeled river showing "omission error"; (b) Agreement in nonriver in modeled and observed river indicate "correct no river"; (c) Representing river in modeled not present in observed indicate "commission error" and (d) Representing agreement in river in both modeled and observed is "correct river".

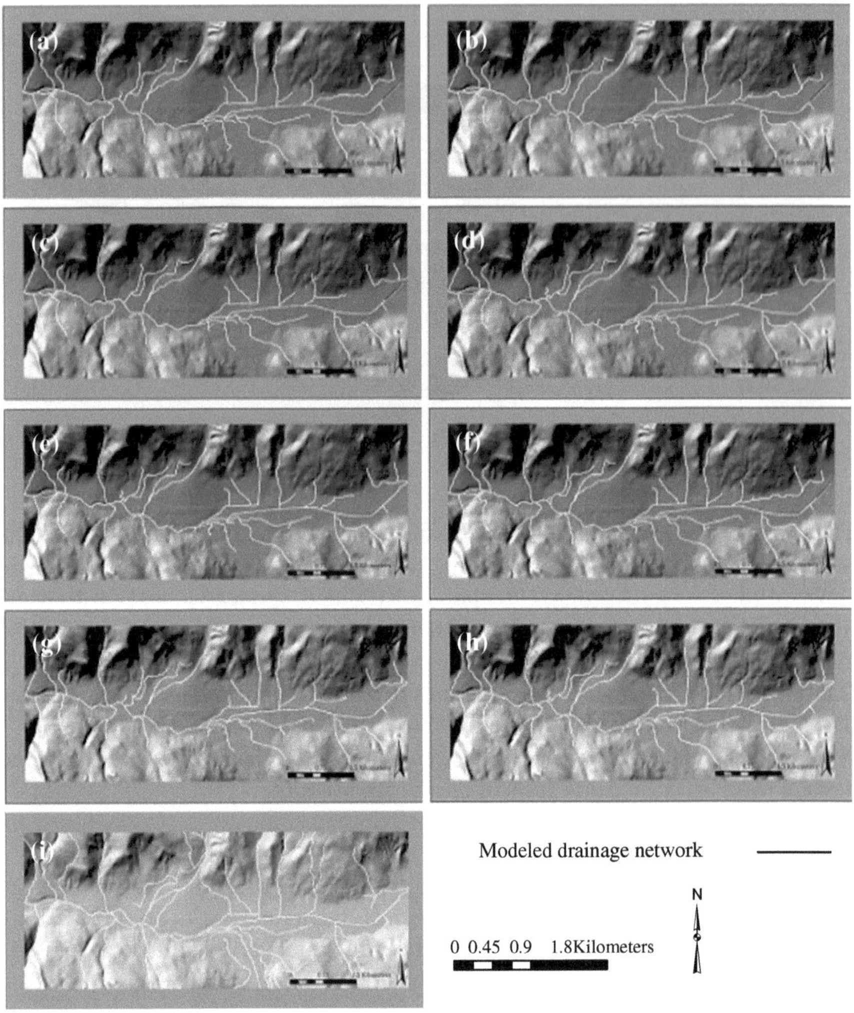

Fig. 6.10 Comparison between the modeled and the reference stream network for **a** original DEM, **b** IDW, **c** Topo to Raster (ANUDEM), **d** Spline, **e** Kriging, Circular, **f** Kriging, Exponential, **g** Kriging, Gaussian, **h** Kriging, Stable, **i** Kriging, Spherical (*Black line* observed; *Red line* modeled)

Table 6.7 Error matrix resulting from the comparison between modeled and observed drainage networks

	Modeled	
Observed	River	No river
River	Correct river	Omission error
No river	Commission error	Correct no river

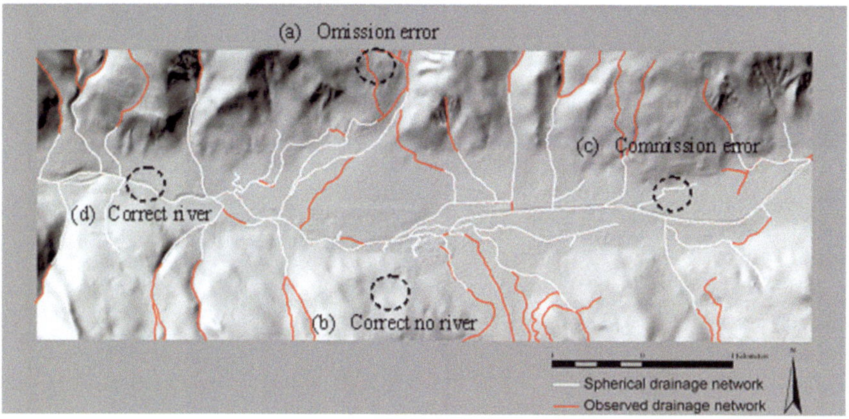

Fig. 6.11 The comparison between modeled and observed drainage networks, Spherical method

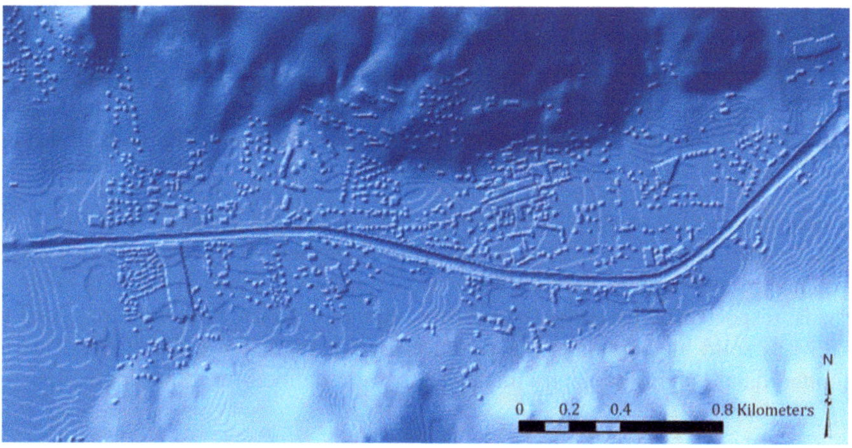

Fig. 6.12 Final DEM (City part) applied in modeling

6.3 Adding Feature Surface to DEM

For flood modeling, the DEM must contain all relevant surface features, such as dykes, that affect the flow of water.

For this purpose, it was necessary to integrate the man-made terrain in the DEM, including roads, dykes and buildings, and their respective values were obtained from the fieldwork and previous databases and research in the area. The final DEM was used in hydrodynamic modeling (Fig. 6.12).

Two separate DEMs were produced: one for the 1957 flood event before reconstruction and the height increase of the dykes, and one after the 1957 flood

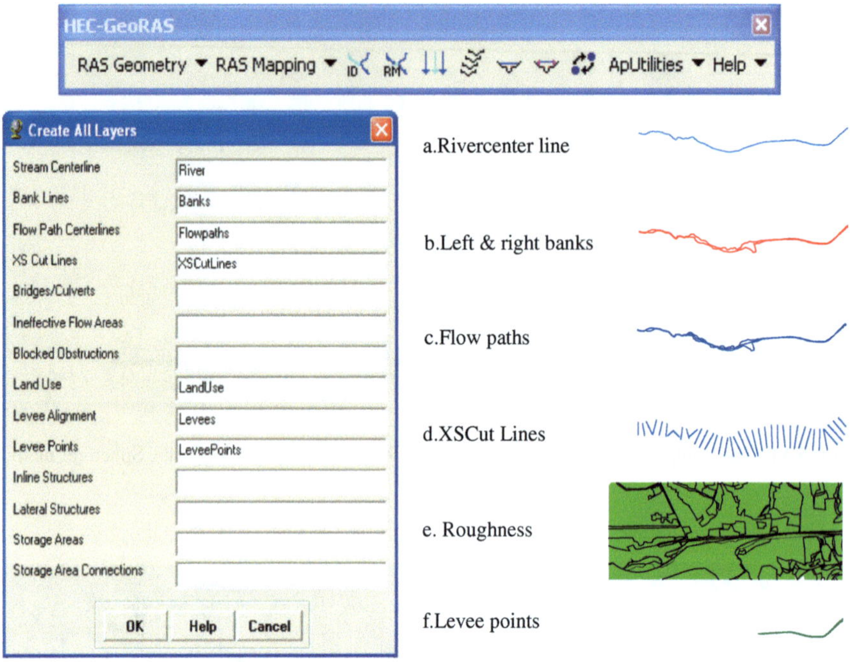

Fig. 6.13 Typical RAS-Geometric data in HEC-GeoRAS applied in SOBEK and HEC-RAS

event when the government decided to increase the height of the dyke about 0.5 m. Road generalization was made in the form of primary, secondary, service roads and tracks, and heights of 0.5, 0.3, 0.2 and 1 m were assigned to different kinds of roads, respectively, based on the literature review [1]. The building also added to DEM as solid blocks hinder the flow water as it surpassed the building. Due to limited data, heights of approximately 3 m were defined for all buildings.

6.4 RAS-Geometric Data

RAS Layers created and edited in HEC-GeoRAS model are necessary in both SOBEK and HEC-RAS models for pre-processing geometric data development and extraction. The layers are included center line, Flow paths, banks, polygon layer of land use to estimate Manning's values, cross sections cut lines, and levees (Fig. 6.13).

After creating and editing each RAS layer, the RAS Geometry data were imported directly into HEC-RAS model as SDF and XML format. These data also imported to in SOBEK model as Ascii and Shp. files.

6.5 Creating Cross Section in the HEC-GeoRAS Model

The extent of cross section locations was represented by the cross sectional cut lines layer based on a 5 m resolution DEM using the HEC-GeoRAS model (Fig. 6.14). As far as if any modeling represents prefect results, these cross sections also needed to be manually modified before being used in flood simulations (SOBEK and HEC-RAS).

In SOBEK, the trapezium method was used to create a cross section and the values of maximum flow width and bottom width and other information were

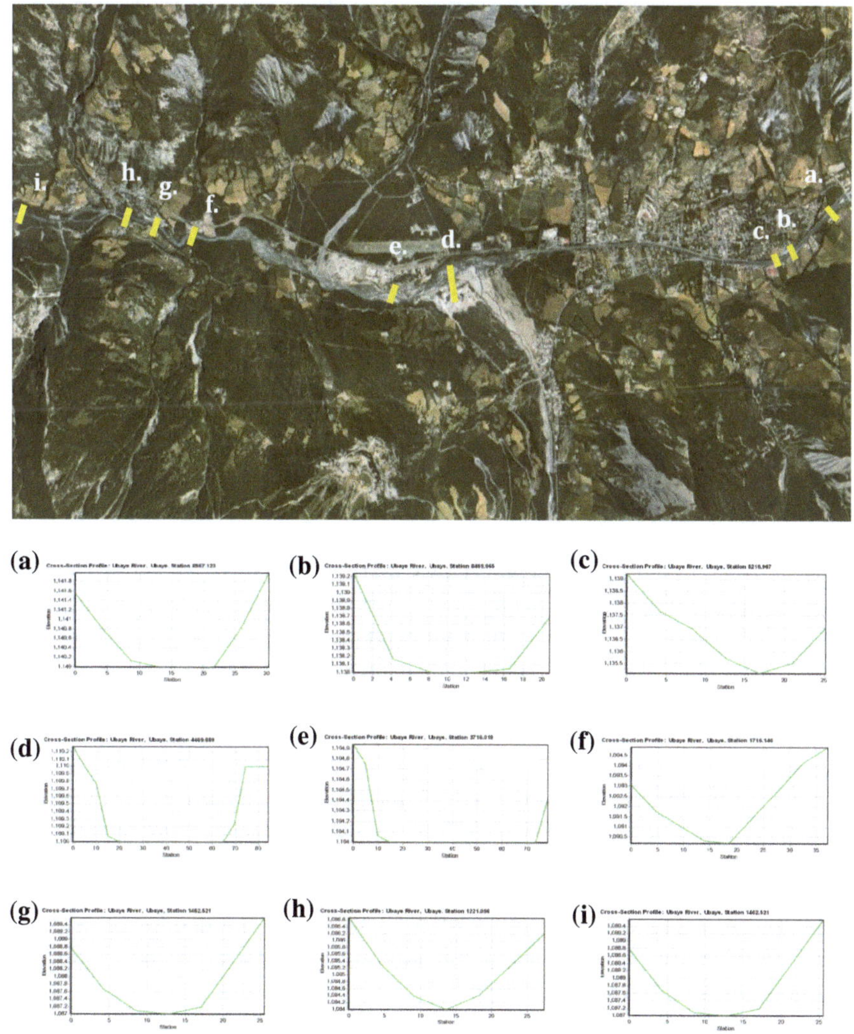

Fig. 6.14 Modeled cross sections at some sampled locations

Table 6.8 The detailed information about selected cross sections and their locations (unit: m)

ID	Shape length	Station	Left bank	Right bank	L-length	C-length	R-length
a	30.331	8967.123	0.075	0.835	499.36	498.80	499.96
b	20.801	8469.964	0.111	0.803	505.91	502.32	495.81
c	25.206	8216.966	0.184	0.876	248.37	246.56	244.49
d	84.362	4469.889	0.072	0.886	4224.3	3754.5	3866.2
e	78.260	3716.019	0.029	0.985	794.70	751.73	788.14
f	37.197	1716.146	0.022	0.916	2056.4	2001.6	2057.2
g	25.528	1462.520	0.063	0.872	232.54	241.46	247.31
h	27.313	1221.056	0.004	0.870	249.64	247.15	246.96
i	28.784	1.52419	0.024	0.960	969.17	973.90	980.11

Shape length represents the length of each cross section; Station represents the distance of each cross section from the upstream; Left bank and Right bank represent the length of cross sections between both left and right sides; L-length represents the distance between two cross sections at the left points; C-length represents the distance between two cross sections at the center points; R-length represents the distance between two cross sections at the right points

defined for each cross section. In HEC-RAS, it was needed to play with the height of the dykes–the left and right bank of cross sections calibrated the model based on the observed data. Table 6.8 represents some necessary information related to each cross section. For example, Object ID (2) represents the second cross section which is located upstream at station 8469.9648. The L-length shows the distance between the second cross section and the previous one on the left side. C-length shows the distance between the second cross section and the previous one in the channel, and R-length shows the distance between the second cross section and the previous one on the right side, and so on.

Ubaye River is divided into three parts: (1) Upstream or channelized part of the river in city area; (2) Middle part of the river in natural environment when the river starts to get wider; (3) Downstream part of the river in natural environment when the river starts to get narrow again. Cross sections should be added at locations where changes occur in slope, width, and roughness. Therefore, to select the cross section's location with respect to all mentioned parameters, 10 cross sections were applied along the river and the data were manually imported to SOBEK model.

Applying more than 10 cross sections in SOBEK made the modeling simulation unstable. The interval between cross sections was selected 150 m, with a total river section of 10 km in length. The cross sections which were created in HEC-Geo-RAS were directly imported to HEC-RAS. In HEC-RAS, there was the possibility to interpolate the cross sections every 20 m. In order to create a more reliable flood modeling result (2D) in HEC-RAS, the extension of the cross sections in the floodplain area was considered mandatory.

As shown in Fig. 6.15, extension of cross section in the floodplain made the model unstable. To prevent such as this kind of error it was needed to modified dykes and roughness values for each cross section manually.

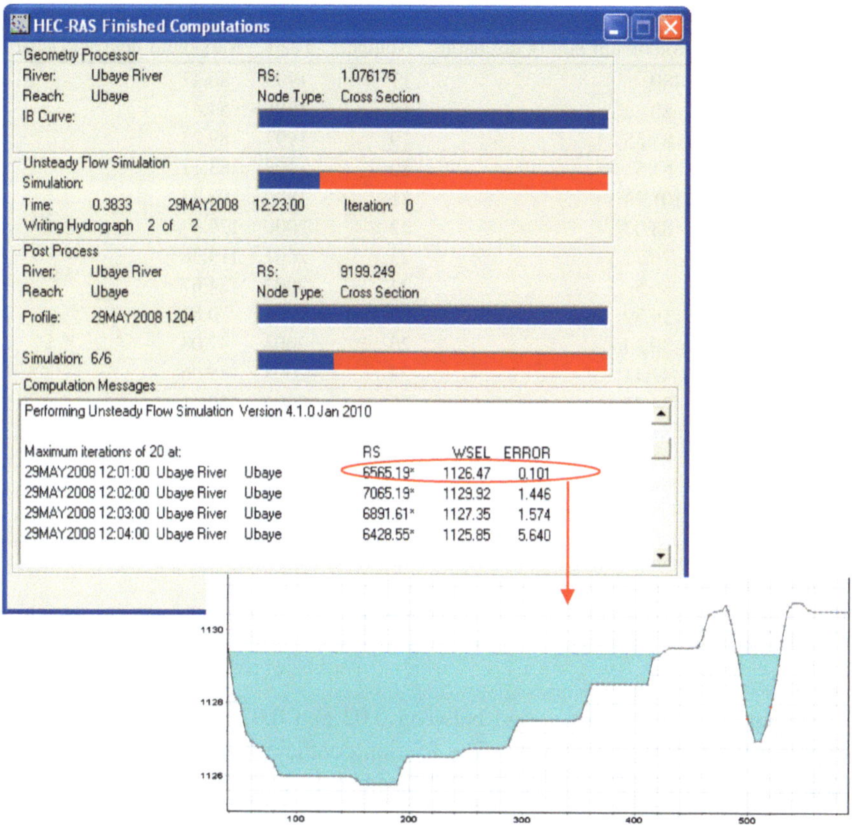

Fig. 6.15 A sample of modeling un-stable due to wrong cross section application

6.6 Discharge Data Modeling

For modeling the discharge value, the rating curve between observed hourly discharge (m^3/s) and hourly water height (m) was plotted to calculate the missing hourly discharge values. The oldest severe flood occurred on 14 May 1957 with a maximum discharge of 480 m^3/s and the nearest flood occurred between 29 and 30 May 2008, with a maximum discharge of 205 m^3/s. Based on the available data, the modeled discharge values showed maximum discharge of 480, 61.5, 80.47, 129.34, and 205 m^3/s for the years of 1957, 1982, 1995, 2000, and 2008, respectively.

Table 6.9 shows the record of maximum discharge of the Ubaye River from 1980 to 2009. But due to the occurrence of an extreme flood event in 1957 and the availability of the data for this event was considered to estimate flood frequency analysis.

Table 6.9 The recorded and modeled maximum discharge (m^3/s) in the Ubaye River

Number	Years	Maximum hourly discharge	Number	Years	Maximum hourly discharge
1	1957[a]	480	17	1995	80.47
2	1980	80	18	1996	54.47
3	1981	87.35	19	1997	89.25
4	1982	61.5	20	1998	83.17
5	1983	101.24	21	1999	92.33
6	1984	88.03	22	2000	129.34
7	1985	76.77	23	2001	119.38
8	1986	86.5	24	2002	84.67
9	1987	59.39	25	2003	70.83
10	1988	30.34	26	2004	42.08
11	1989	42.41	27	2005	89.06
12	1990	34.75	28	2006	50.41
13	1991	81.51	29	2007	18.72
14	1992	65.18	30	2008[a]	205
15	1993	86.37	31	2009	89.07
16	1994	182.92			

[a] The recorded maximum discharge (m^3 /s)

6.7 Roughness Values

The roughness coefficients varied between 0.02 and 0.07 $m^{-1}/3^{s-1}$ according to characteristics of natural channel. The friction coefficients were adjusted in order to achieve the best fit between the simulated and observed water stages. These roughness values are based on land use maps which were digitized according to related aerial photos (Fig. 6.16).

Based on the available aerial photo, six roughness maps were prepared to use as an important input data in flood modeling. The results of the coefficient values are presented in Fig. 6.17. Distribution of roughness coefficients in a channel section is an important aspect that needs to be addressed properly. A roughness value underestimates the discharge and a low value can overestimate.

The selection of a suitable value of n is the single most important parameter for the proper estimation of velocity in an open channel.

For being sure about the correct selected roughness value for the channel, the model was run for different values of 0.02–0.07. The result of hydraulic depth was compared with observed data at Abattoir Bridge. Manning's value of 0.02 was

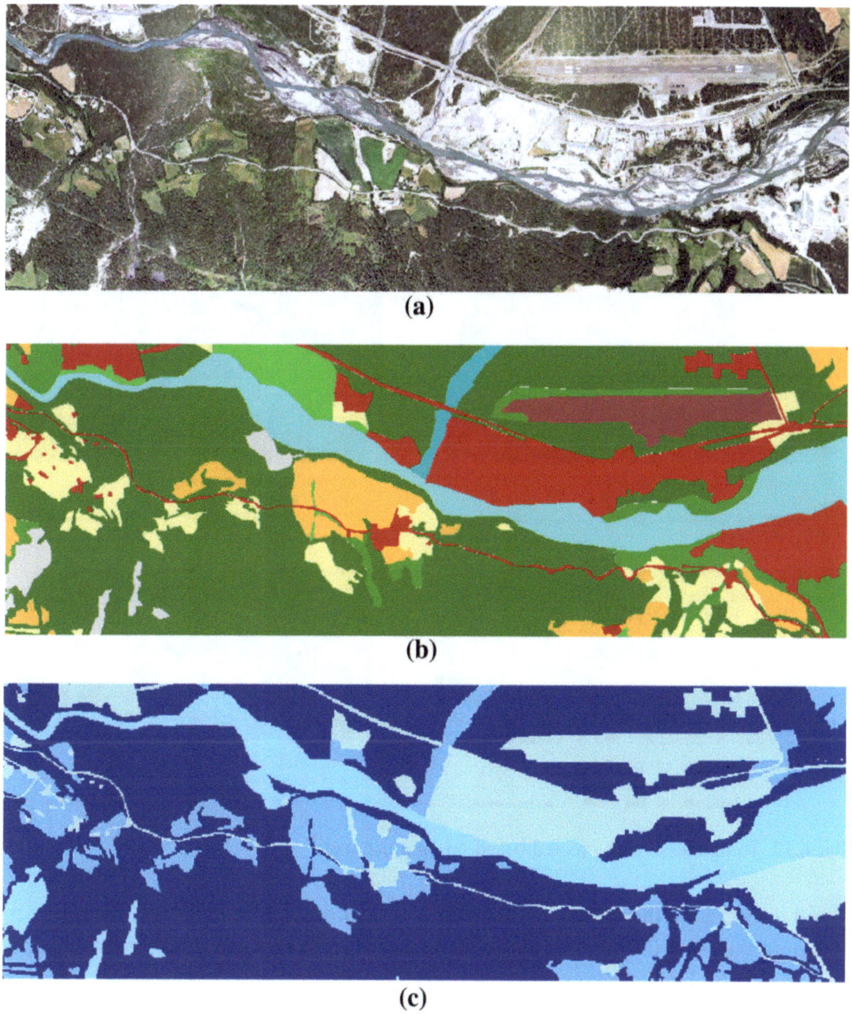

Fig. 6.16 The model procedure to create roughness map. **a** Available aerial photo. **b** Land use map created and digitized in ArcGIS. **c** The roughness map in raster format

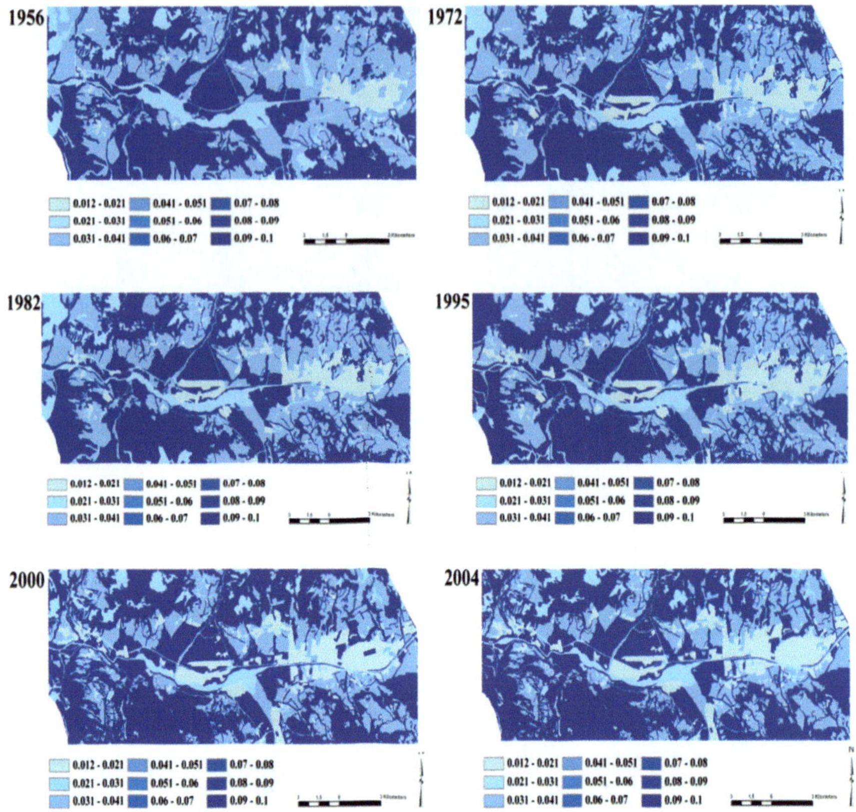

Fig. 6.17 Roughness value created based on land use maps

found to decrease the hydraulic depth in the channel and both banks, inversely the Manning value of 0.07 increased the hydraulic depth of water for narrow channels in city part of Ubaye River.

As shown in Fig. 6.18, applying the hydraulic depth did not represent a correct agreement with observed data in Abattoir Bridge. These results confirm that the roughness coefficient has an important role in flow behavior. By implementing the right value, the appropriate result was obtained which could construct the real event of 2008 in the area. The calibration procedure based on roughness values is discussed in this section.

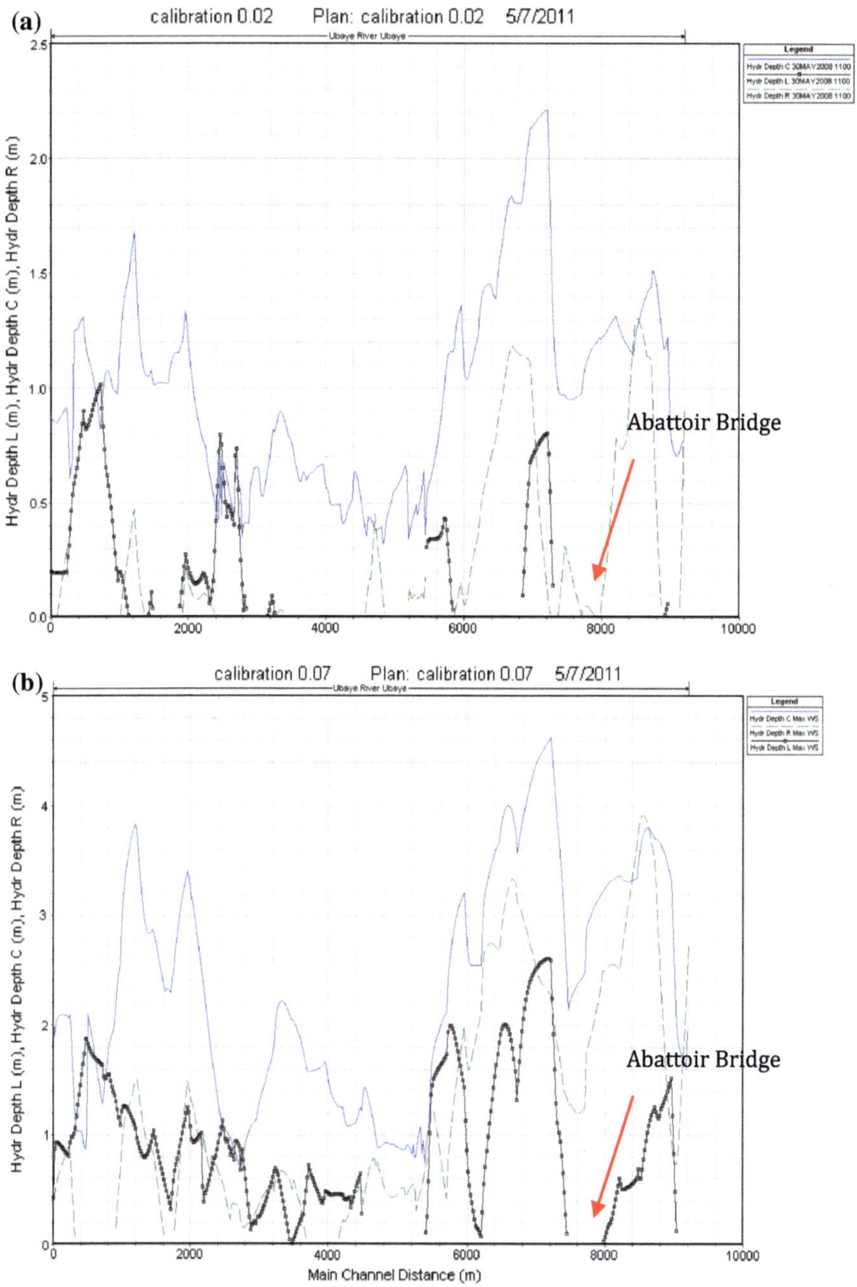

Fig. 6.18 Applying two different Manning's coefficient values for the channel. **a** The value of 0.02. **b** The value of 0.07

6.8 Calibration Procedures

In this study, due to a lack of available recorded data over the case study area, calibration analysis was limited to the observed height of water levels found in the literature and field reports of the 2008 flood. The adjustment was performed using different friction values in the channel. Table 6.10 shows the comparison between observed and modeled values for water depth. The results are also plotted as shown in Fig. 6.19.

The results for different test runs based on different friction values to analyze the sensitivity of the model are shown by comparative graphs of observed values and simulated values of water depth. Calibration of the model was also done based on this sensitive element of friction value. The values correspond to the horizontal and vertical friction values in the channel.

Model sensitivity analysis of friction values was performed based on changing Manning's coefficient values of roughness within the channel. The surface roughness map for different land use types was generated for overland flow modules. The results of testing show that the friction value of 0.05 in the channel represents the best agreement between modeled and observed water height

Table 6.10 Comparison between observed and modeled depth of water (m)

Observed depth	Modeled (0.03)	Modeled (0.06)	Modeled (0.05)	Modeled (0.04)	Modeled (0.07)	Modeled (0.02)
1.39	0.78	1.13	1.03	0.91	1.23	0.61
1.46	1.02	1.47	1.34	1.19	1.74	0.89
1.49	1.08	1.55	1.42	1.26	1.78	0.91
1.54	1.1	1.58	1.44	1.28	1.84	0.95
1.59	1.14	1.64	1.5	1.34	1.9	0.98
1.67	1.18	1.7	1.55	1.39	2	1.04
1.66	1.23	1.76	1.61	1.44	2	1.04
1.68	1.24	1.77	1.62	1.45	2.02	1.05
1.65	1.25	1.79	1.64	1.46	1.98	1.03
1.67	1.26	1.79	1.64	1.47	2	1.04
1.7	1.36	1.93	1.77	1.58	2.16	1.13
1.85	1.4	1.98	1.82	1.62	2.22	1.17
1.94	1.47	2.08	1.91	1.71	2.33	1.23
2.02	1.53	2.17	2	1.78	2.42	1.28
2.04	1.55	2.19	2.02	1.8	2.45	1.3
2.04	1.55	2.19	2.02	1.8	2.45	1.3
2.07	1.57	2.22	2.05	1.83	2.48	1.32
2.22	1.72	2.41	2.22	1.99	2.69	1.43
2.16	1.65	2.32	2.14	1.91	2.61	1.39
2.24	1.64	2.31	2.13	1.9	2.73	1.44
2.17	1.63	2.3	2.11	1.89	2.62	1.4
2.23	1.63	2.29	2.11	1.88	2.57	1.37
2.16	1.45	2.05	1.88	1.69	2.21	1.16
2.17	0.78	1.13	1.03	0.91	1.79	0.92

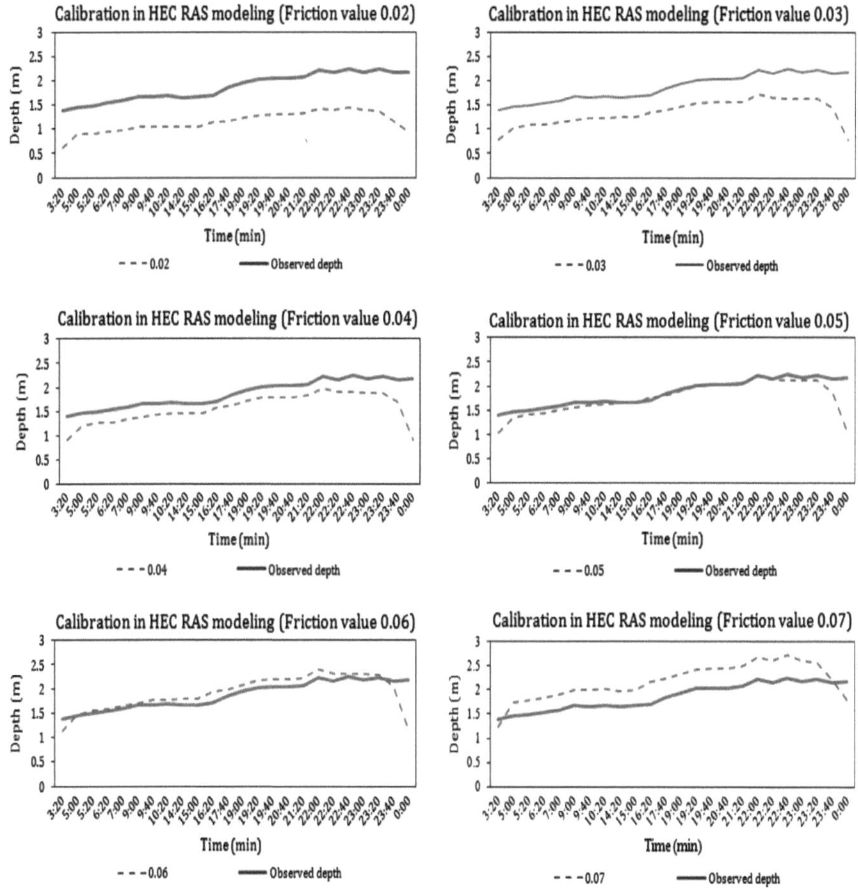

Fig. 6.19 Calibration based on Manning's coefficient values

measured close to the Abattoir Bridge. Because of a lack of data the calibration was only based on one point measurement data. More observation points within reach could have allowed for flood routing, leading to better calibration of the model. Taking into account the length of the river and lack of observed data, the model was considered to have been successfully calibrated. Manual calibration of the 1D model was employed to constrain friction parameters within a feasible range. Since precise water level predictions are crucial for dyke breach modeling, calibration against distributed observations was preferred over the model, fit to discharge series at the measurement gage. Figure 6.20 represents the Q–Q plot of regression standardized residual for different friction variables.

Statistically, the best result in terms of R, R-square, and standard deviation error was belonged to friction values of 0.05 (Table 6.11).

The 2008 event was selected because the observed hourly water height was available. In this research, although there was limited observed data, the results

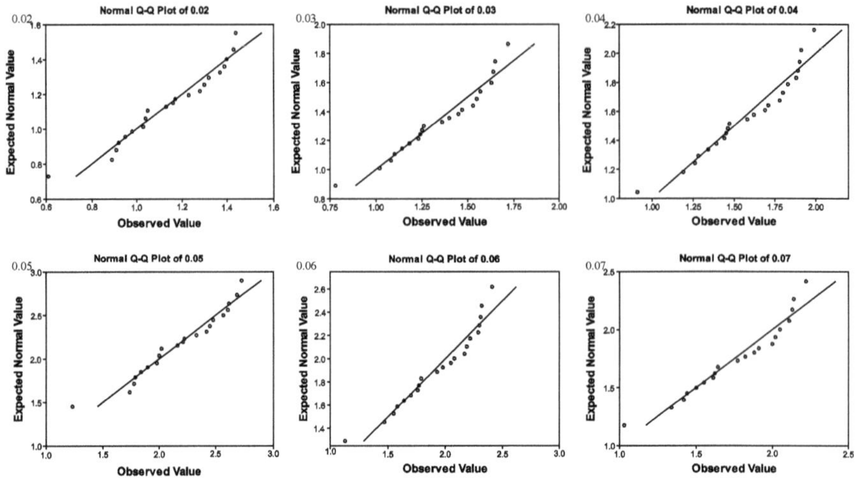

Fig. 6.20 Q–Q plot of regression standardized residual for different Manning's coefficient values

Table 6.11 Statistical analysis for different Manning's coefficient values

	0.02	0.03	0.04	0.05	0.06	0.07
R	.844	.735	.730	.845	.729	.732
R-square	.712	.540	.533	.715	.531	.535
Standard error	.203	.186	.215	.115	.256	.238

Table 6.12 PF and OF function for different Manning's coefficient values

	0.02	0.03	0.04	0.05	0.06	0.07
Station	1	1	1	1	1	1
Time (hour)	24	24	24	24	24	24
Observed height water (m)	1.7	1.7	1.7	1.7	1.7	1.7
Simulated discharge (m^3/s)	17.38	28.15	41.6	94.39	70.18	55.95
Simulated height water (m)	1.13	1.36	1.58	1.77	1.93	2.16
Penalty function	7.2	4.1	2.4	0.5	1.0	1.6
Object function	31.8	28.4	26.6	24.4	24.8	25.1

show that the error and level of uncertainty was at the lowest level. The results of penalty function are shown in Table 6.12.

Based on the results, the best agreement between observed and modeled height of water was gained when the friction value of 0.05 was used for channel. The optimization method was selected based on low object function.

In calibration procedure, all different friction values were applied in HEC-RAS model simulation, and depth of water for different values was compared. The successful modeling as expected before was the modeling with the friction value of 0.05 for river channel.

6.9 Hydrodynamic Modeling

River hydraulic investigations and studies include the elevation of flow characteristics and physical behavior of rivers, e.g., prediction of stage, discharge, velocity, shear stress, and so on. Basic hydrographic survey data are a critical component of these studies. In order to build up the hydrodynamic model, a series of input data were needed such as the DEM, cross sections, dyke position and height, aerial photos for digitizing the land use maps, rating curve, flood hydrograph, and roughness coefficient for different types of land use.

The variability of flood crisis scenarios such as DEM scenario (different resolution of 5 and 10 m), river change scenario, dyke failure scenario, and model scenario was estimated for civil protection purposes in the study area, with respect surface elements which affect water flow, a variety of cross sections and river morphology over time.

A selection of a subset of these scenarios with distinct return periods was applied as input information into the hydraulic (1D) and hydrologic (1D–2D) models. The results of hydrodynamic modeling are presented in this chapter and they are discussed in details.

6.10 Scenarios

6.10.1 DEM Scenario

The work described here aims to compare the different DEM resolutions and their effects on reality and on the accuracy of the results for the purpose of flood modeling. Within this, a method entitled "Resample technique" was also applied to change the resolution of the available DEM from 10 to 5 m. The idea behind this was to estimate how much the resampled data are close to the data source.

The calculated water velocity by SOBEK and HEC-RAS models shows that this hydraulic parameter decreases with increasing DEM pixel size. This is related to the fact that the model considers more roughness and friction as it uses the higher resolution DEM. A major disadvantage of using the low resolution DEM is the loss of some features such as dykes and levees. These features play an important role on flood model simulation results.

In flood inundation modeling, DEMs influence the model outputs. Drainage networks used as input datasets in such models are derived from DEM by connecting each cell to its neighbor in the direction of the principle slope. By comparing the results, flat surfaces are typically the result of an inadequate vertical DEM resolution, which can be further worsened by a lack of horizontal 10 m resolutions. With the increase of horizontal and vertical resolutions, delineated river networks from DEM become closer to the actual.

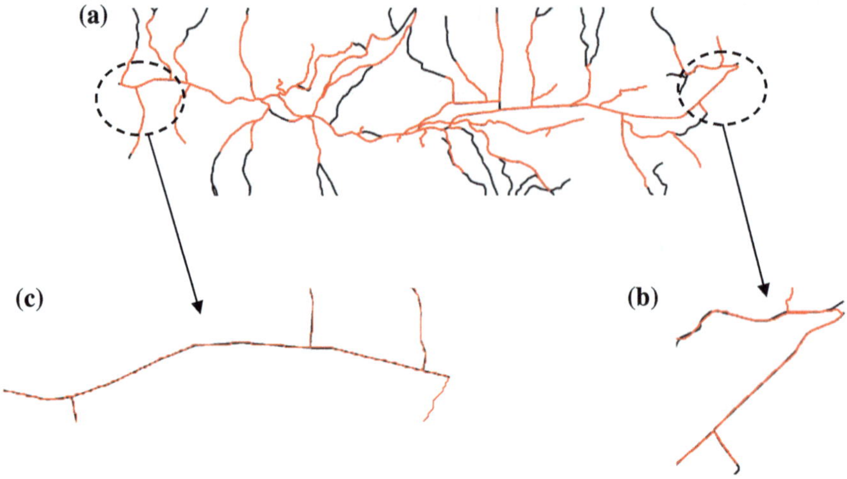

Fig. 6.21 **a** Generated river network from 5 m resolution DEM (*red color*) overlapped on actual river network observed in the field (*black color*); **b** the agreement between observed and modeled in upstream; **c** the agreement between observed and modeled in downstream

Figures 6.21 and 6.22 show the delineated river network from 5 and 10 m resolutions which were compared with actual river networks observed in the field. Visualization analysis shows that due to a low resolution of 10 m, compared to 5 m, the generated river network for 10 m deviates more and does not agree well with the actual drainage paths. But with the increase of horizontal and vertical

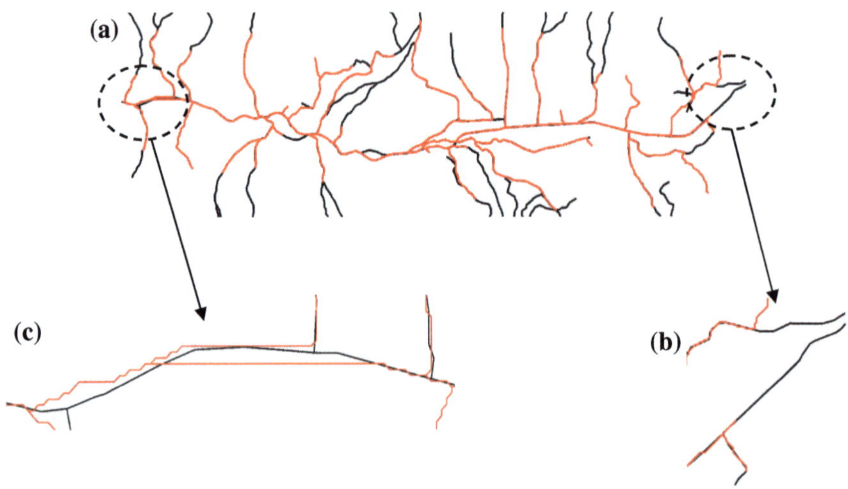

Fig. 6.22 **a** Generated river network from 10 m resolution DEM (*red color*) overlapped on actual river network observed in the field (*black color*); **b** the agreement between observed and modeled in upstream; **c** the agreement between observed and modeled in downstream

resolutions, the delineated river networks from 5 m DEM come in close agreement to the actual flow path. Even in flat urban areas, the results of 5 m resolutions are more satisfactory compared to 10 m resolutions.

To evaluate the quality of two available DEMs with different resolutions and to calibrate the models, the recent recorded flood event in May 2008 was considered as a reference. In the 2008 flood, the water flow remained between the channels and there was no inundation in the floodplain. The flood event was simulated twice by models, first with the 5 m resolution DEM and second with the 10 m resolution DEM. For both simulations, the remaining input data stayed the same. Figure 6.23 shows the simulated maximum flood inundation maps for both models. It can be observed that the simulated flood for the higher resolution agrees well with the 2008 event, based on recorded field measurements. When applying the 10 m resolution DEM, the inundation areas were delineated and over topping flow also occurred. On the contrary, there were no inundated areas in both SOBEK and HEC-RAS models using the 5 m resolution DEM.

Therefore, it could be concluded that a 5 m resolution DEM shows flood flow paths more accurately and in more detail in comparison with the 10 m DEM

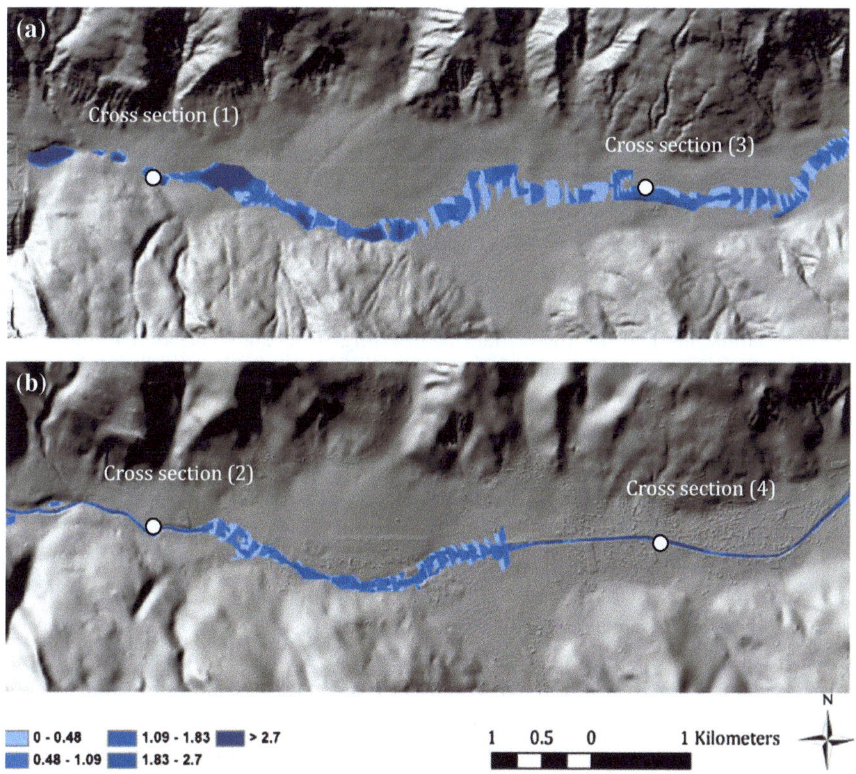

Fig. 6.23 Comparison of flood simulation (depth of water). **a** DEM with 10 m resolution, **b** DEM with 5 m resolution

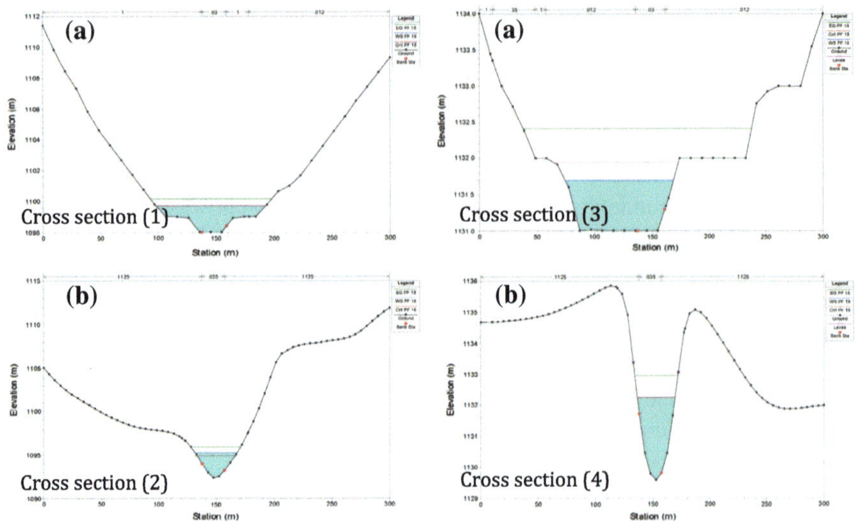

Fig. 6.24 Comparison of river cross sections at two different points. **a** DEM with 10 m resolution; **b** DEM with 5 m resolution

resolution. This may be due to the fact that a 10 m resolution DEM ignores and generalizes surface features which affect the water flow. Therefore, the preferential water paths are not well defined. The results of the 10 m DEM simulation show that the flow distribution shifted from the actual flow paths in locations of the generated river network. It was obvious that the flood mainly occurred due to overflow in the rivers, which was perhaps due to the negligence of the locally elevated lands such as dykes and roads in the model. In the real situation, these elevated lands blocked the movement of flooding from the riverside to other side. However, in this simulation process, the 10 m resolution DEM could not capture such local elevations in a few meter widths, and thus the simulation results were affected. Figure 6.24 represents the cross section at two different locations in upstream and downstream, with different DEM resolutions.

Figure 6.25 represents the hydraulic water depth for DEMs with 5 and 10 m resolutions. The hydraulic depth was compared with the results of 10 m resolution DEM. Modeling of the 2008 flood event had a maximum discharge of 205 m^3/s. The model indicated that maximum water depth and velocity in the Ubaye River were approximately 2 m and 3 m/s, respectively. This corresponds to the field measurement reports of the 2008 flood. As a consequence, it could be concluded that 5 m resolution DEM shows a successful simulation and has the best agreement with the observed recorded data.

As mentioned before, the second aim in this section was to compare resampling techniques in changing DEM resolutions. The intention was to resample the low resolution DEM to a high resolution and to compare the low resolution DEM before and after resampling it with the original high resolution DEM.

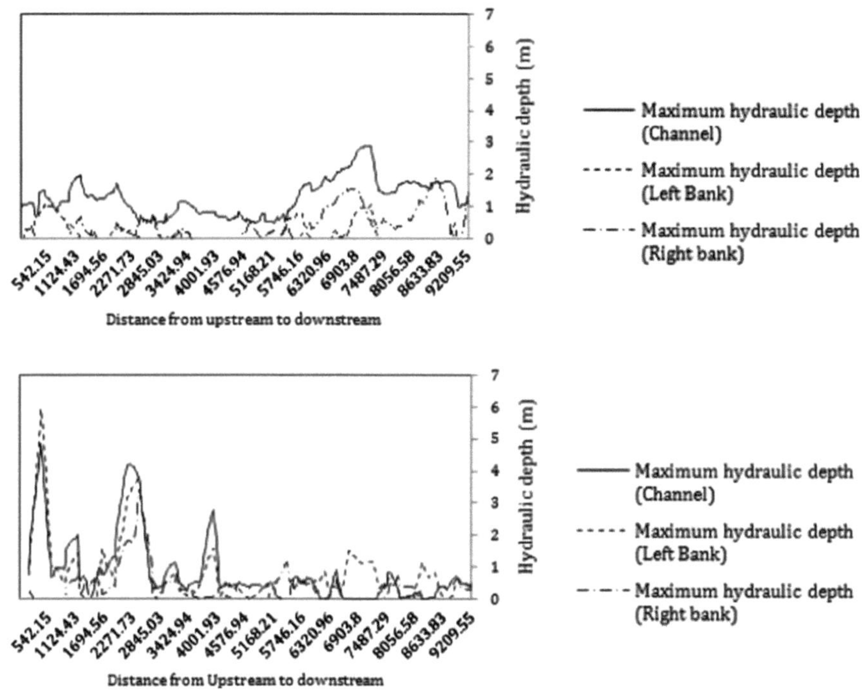

Fig. 6.25 Hydraulic depth of water for **a** DEM 5 m resolution; **b** DEM 10 m resolution

Table 6.13 *R*-square and RMSE between DEM 10 m resolution before and after resampling with high resolution DEM 5 m

	R-square	Mean	Max	RMSE
DEM 10 m before resampling	.894	25.928	322.279	41.284
DEM 10 m after resampling	.995	4.292	155.933	8.966

The production of a large number of elevation error estimates, evenly distributed over the DEM, makes it possible to explore some general questions such as, how to identify the statistical characteristics of DEM error, and how much this differs between the resampled method and the relationship between RMSE of elevation.

The selection of an optimal set of parameters to be used in resampled DEM required the exploration of different combinations of values. Table 6.13 shows four statistical analyses for each DEM before and after resampling, comparing it with high resolution DEM (5 m) as a reference. The statistical analysis is based on the following parameters:

1. RMSE calculated for selected data points (randomly 10,000 points)
2. Mean error presents the arithmetic mean of the error values and reveals whether the interpolation has a tendency to under or overestimate on average.
3. Maximum absolute error shows how large the errors can be in extreme cases

Table 6.14 SAM statistical results for comparing estimated elevation with high resolution DEM (5 m)

	DEM 5 m	Estimated	Residuals
DEM 10 m before resampling			
Min	1069	1120.773	−352
Max	1749	1615.079	527.936
Mean	1245.237	1245.237	<.001
S.D.	126.646	91.439	87.625
DEM 10 m after resampling			
Min	1069	1073.306	−135.696
Max	1749	1756.126	157.32
Mean	1245.237	1245.237	<.001
S.D.	126.646	126.311	9.206

4. *R*-square is a statistical measure of how well a regression line approximates real data points; an *R*-squared of "1.0" (100 %) indicates a perfect fit.

As a general conclusion, in terms of RMSE analysis, better results in the accuracy of the model were obtained when a DEM was used after resampling. The mean error in resampled DEM was 4.292, much less than the value of 25.928 for DEM before resampling. Maximum absolute error also shows that the errors are not so large in DEM after resampling. *R*-square value between the regression line of both DEMs shows that, in terms of *R*-square, the results fitted well in DEM especially after resampling procedures. Therefore, as an overall result, it could be concluded that a low resolution DEM could show better and more accurate results when it is resampled to a high resolution DEM.

For other additional analysis a Geostatistical based software so-called "SAM" was applied. It offers a comprehensive array of spatial statistical methods. The methods available in SAM range from simple tools for exploratory graphical analysis (e.g., mapping and graphing) and descriptive statistics of spatial patterns (e.g., autocorrelation metrics), to advanced spatial regression models (e.g., autoregression and eigenvector filtering). Table 6.9 represents the SAM statistical analysis for comparing estimated elevation with high resolution DEM (5 m).

First, the analytical outcome in "SAM" environment was the standard regression output. According to the results, Standard error for DEM 10 m before resampling is 0.007 and after resampling this value decrease to <.001. The *P* value in both models is <0.001. Second, the predicted values for each method were compared with original values of source data points. The minimum, maximum, mean, and standard deviations for each method were calculated and the descriptive statistical results are shown in Table 6.14. These parameters are defined as following:

- Minimum absolute error shows how small the errors can be in case of considering lowest elevation in both observed and estimated value.
- Maximum absolute error shows how large the errors can be in case of considering highest elevation in both observed and estimated value.

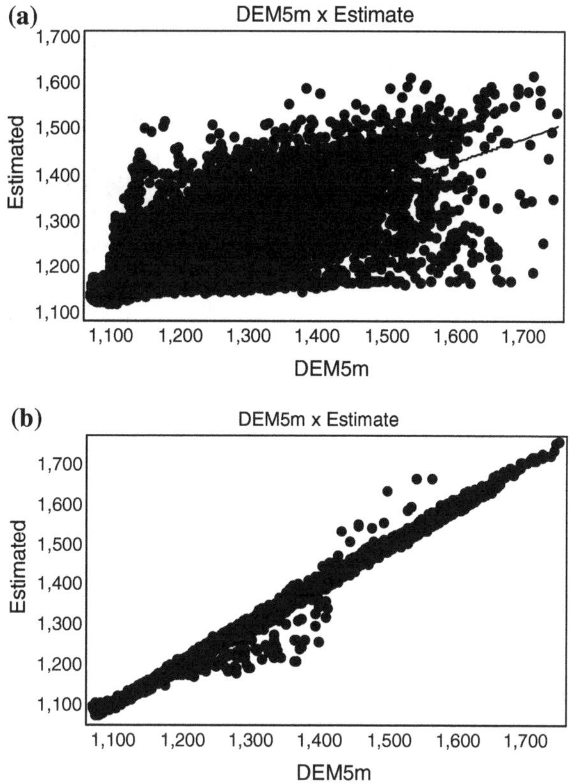

Fig. 6.26 Linear regression analysis for estimated values. **a** Comparison between DEM 5 m and DEM 10 m before resampling; **b** comparison between DEM 5 m and DEM 10 m after resampling

- Mean error presents the arithmetic mean of the error values and reveals whether the interpolation has a tendency to under or overestimate on average.
- Standard deviation shows how much variation or "dispersion" there is from the average. A low standard deviation indicates that the data points tend to be very close to the mean, whereas high standard deviation indicates that the data points are spread out over a large range of values.
- "Residuals" are defined as meaning the spatially modeled component of variation not accounted for by the environmental variables, and that is why it has a very strong spatial structure (because it is added when modeling the semi-variogram). By doing this, it is 'released', at the same time, the effect of the spatial structure.

As shown in the Table 6.14, mean error in both modeling results is p value $<.001$. Other estimation results show that DEM 10 m after resampling was in better agreement with the original high resolution DEM. Furthermore, the estimated values are close to the 5 m DEM resolution values. The next analysis refers to distribution of the estimated and residuals values between DEM 10 m (before and after resampling) and DEM 5 m. Figure 6.26 a and b shows the linear regression analysis which is modeled on the relationship between 10,000 randomly

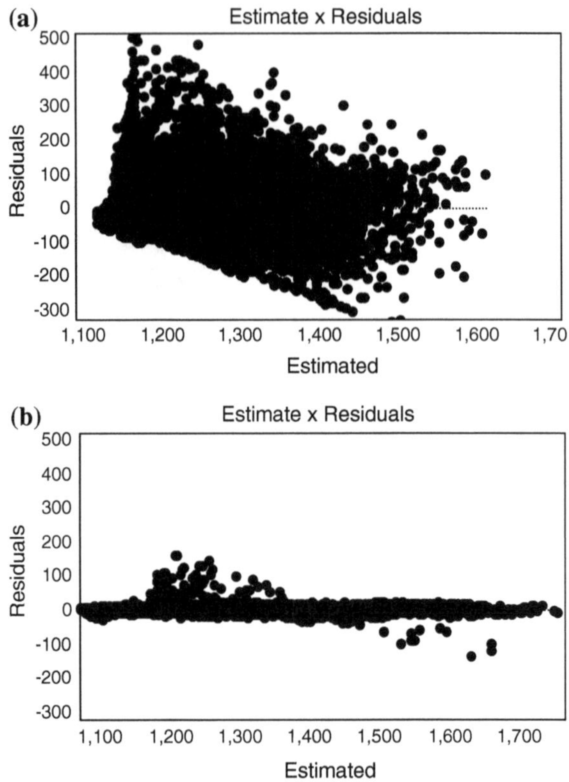

Fig. 6.27 Linear regression analysis for residual values. **a** Comparison between DEM 5 m and DEM 10 m before resampling; **b** comparison resampling DEM 5 m and DEM 10 m after resampling

sampled data points between a dependent variable (DEMs 10 m before (a) and after (b) resampling) and an independent variable (DEM with 5 m resolution). As shown in Fig. 6.26a and b, the estimated and residual values are better fitted with the values of DEM 5 m and a linear regression is statistically significant. Figure 6.27 shows linear regression analysis for residual values. a. Comparison between 5 and 10 m before resampling; b. comparison between 5 and 10 m after resampling.

A further characterization of the data includes skewness and kurtosis. To this propose, the standard residual diagnosis graphs, including the histogram of the residuals was plotted which seems to be normal in both models in Fig. 6.28a and b. The skewness values for DEMs before and after resampling are equal to 1.14 and 4.11. It seems that the skewness for DEM before resampling is symmetry or precisely, because it looks same to the left and right of the center point. Positive values for the skewness indicate data that are skewed right. Skewed right, means that the right tail is long relative to the left tail.

The Kurtosis value for resampled DEM showed high Kurtosis tend to have a distinct peak near the mean, decline rather rapidly, and have heavy tails. Dataset in DEM before resampling indicated low kurtosis tend to have a flat top near the mean rather than a sharp peak (Fig. 6.28).

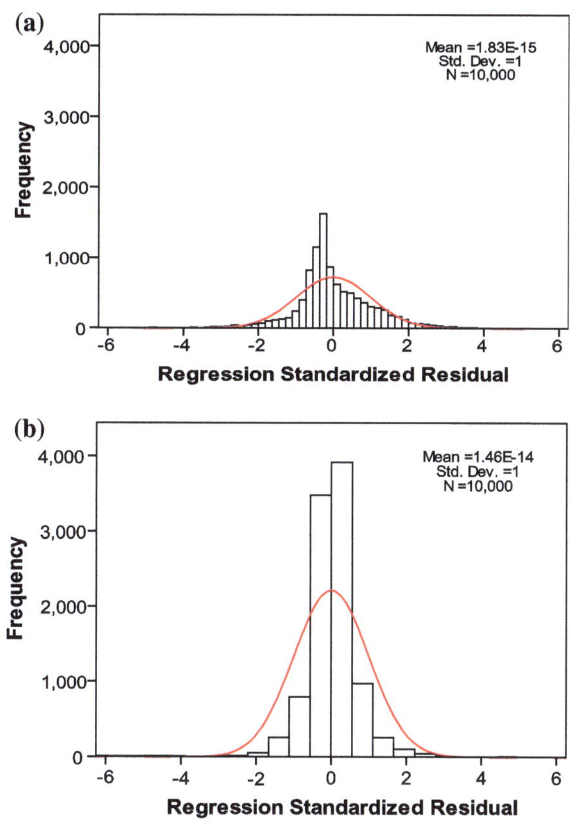

Fig. 6.28 Frequency distribution of DEMs error. **a** Comparison between 5 and 10 m before resampling; **b** comparison between 5 and 10 m after resampling

Figure 6.29 illustrates the results of Table 6.14 as graphs, with the red representing the source values and blue representing estimated values. Perhaps the most interesting screens (in terms of what SAM does compare with most software) are exploring spatial patterns in the model, especially in model residuals (the green line of the correlogram).

Negative (positive) values indicate negative (positive) spatial autocorrelation. Values range from −1 (indicating perfect dispersion) to +1 (perfect correlation). A zero value indicates a random spatial pattern. The high Moran's I in both DEMs before and after resampling showing perfect correlation in the first distance (Table 6.15).

Figure 6.30 shows the comparison between DEM 5 m resolution, DEM 10 m resolution before resampling and DEM 10 m resolution after resampling in schematic shape of the 3D plot for 10,000 selected point elevations.

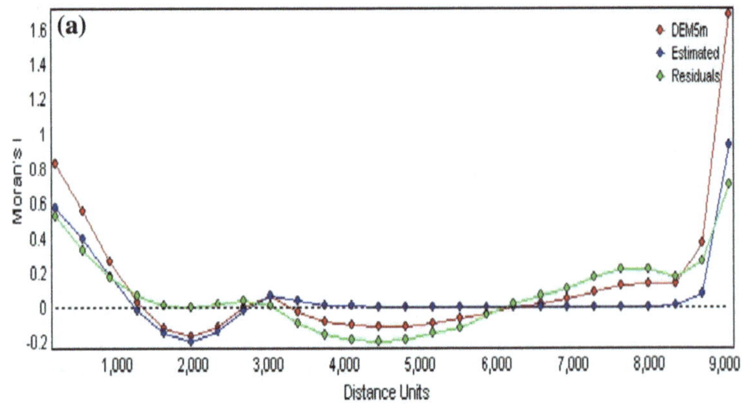

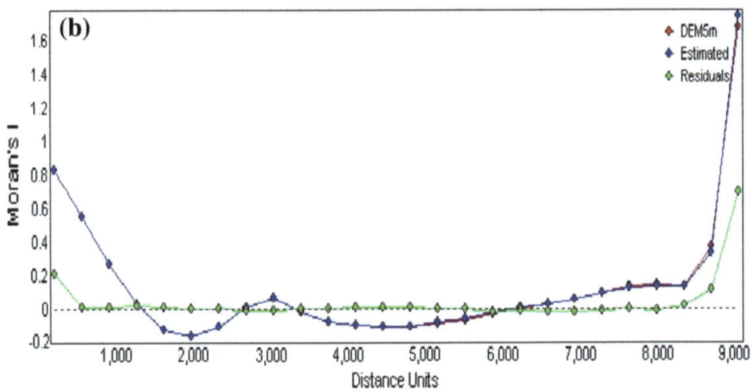

Fig. 6.29 Moran's analysis. **a** Comparison between 5 and 10 m before resampling; **b** comparison between 5 and 10 m after resampling

Table 6.15 The high Moran's *I* in the first distance

Method	Moran's *I* value	Result
DEM 10 m before resampling	0.526	Perfect correlation
DEM 10 m after resampling	0.208	Perfect correlation

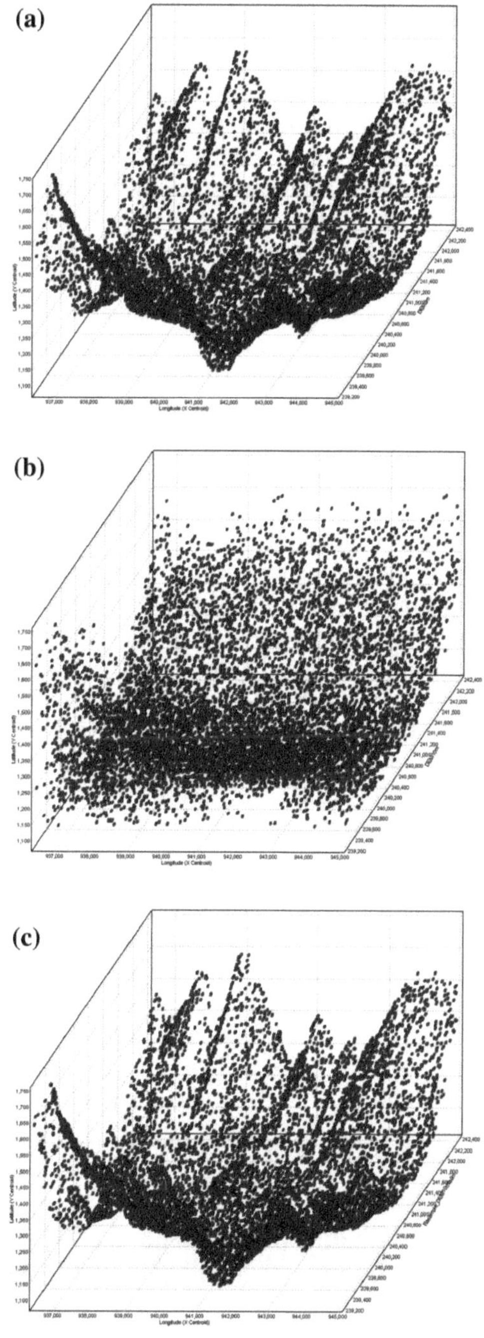

Fig. 6.30 3D plot for 10,000 selected point elevations for **a** DEM 5 m resolution; **b** DEM 10 m resolution before resampling; **c** DEM 10 m resolution after resampling

6.10.2 River Change Scenario

According to the literature and the RTM report, in contrast to the period of 1970–1990, the active channel was characterized by reduced peak flow and vegetative re-colonization (+5 %) of the active channel (Fig. 6.31). The evaluation of the active channels' average width and annual peak flow from 1945 to 1990 on the middle of the Ubaye River is shown in Fig. 6.28.

River change detection was estimated for 1957, 1982, 1995, 2000, and 2008 according to available aerial photos, land use maps and discharge values. The results of this analysis are as follows:

Flood 1957: Due to the lack of access to any data regarding past events, it was not possible to obtain references and increase the accuracy of the results. For instance, the flood event of 1957 has been recorded in every commune that was affected by flood and the consistency in the reports varies in different communes. The 1957 flood occurred on June 14. Various geomorphic phenomena along with climatic factors enhanced the damaged cause by this severe event. Climatic factors such as precipitation, snowmelt, and wind triggered the flood in this area. Other factors such as debris accumulation and particle size varying from mud to several cubic meters also affected this flood event. Based on the available literature, local snow jams and snow melting at an elevation between 2,000 and 2,500 m, followed by approximately 13 showers a day, also allowed the debris studded streams to flow down the valley at a very high speed wherever the slope was moderate to high. The uncertainty behind the 1957 flood simulation has to do with the lack of observed data and information about this event. The data were missing because the equipment was damaged and there was an interruption in reading the data [1].

Based on the limited data in some reports, the maximum discharge of this event was 480 m^3/s. The peak of the hydrograph reaches toward the end of the event

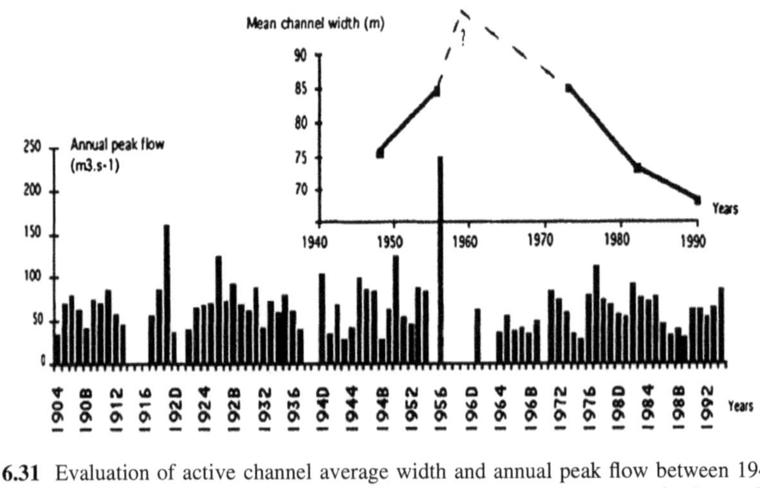

Fig. 6.31 Evaluation of active channel average width and annual peak flow between 1945 and 1990 on the middle Ubaye River (after Hydro bank data and aerial photograph (*Source* RTM)

with a steady increase in the discharge level throughout the period. The initial period of the curve flattening may be attributed to the steady rise in the river water as a result of the water flow constantly accumulating from several streams from the upstream. The high increase of the peak discharge may be associated with sudden snow melt attributed to warm wind from the south, leading to an increase in the water level several folds within a very small period of time.

An essential factor that affected the intensity of the event was the amount of debris that was transported with the running water down the slope and blocked the river flow path [1]. But there were no available data for the quantity of debris accumulation during this event. Based on the assumption that all other factors are constant, only the estimated discharge data were used as input to the model and the changed conditions (e.g., dykes) were incorporated within the DEM. The results of the 1957 flood simulation (Fig. 6.32) were validated based on the photographs available from RTM.

There were no data regarding velocity or depth of water during this event. Therefore, a model calibration for these parameters without any observed data was almost impossible. The 1957 flood is validated at two points with available photographs as shown in the Fig. 6.32. These validations gave at least some clarity about the severity of the flood. The maximum estimated velocity was ~6.32 m/s and the maximum water depth was ~3 m in the channel.

Flood 2008: In May 2008, the Ubaye River exhibited a peak discharge of 205 m^3/s. If no major flooding occurred, some damages were observed: a bridge had been destroyed, and in some specific areas, river ridges have been scored (Fig. 6.33). Based on the RTM, the triggering factor for this event was heavy rainfall and actively melting snow in the upper part of the slopes, a similar situation to the 1957 flood.

During this event, the floodplain was not affected by inundation like the 1957 flood. The water stayed within the banks, but in some places the banks became wet and overtopped. The bridge of "Pont de l'Abattoir" where the measuring station was located had its scale a few centimeters out of the river flow (Figs. 6.34 and 6.35).

The DEM that was used in this simulation was adjusted by increasing the heights of the dyke by 0.5 m from that of the 1957 event. Figure 6.32 represents the model results in the city area of the Ubaye River. The maximum estimated velocity for the 2008 flood in the channel was ~3 and had a maximum water depth of ~2.7. At cross section 12 located exactly at the Abattoir Bridge, the maximum water depth in the near bank was 1.7 m. This result shows the best agreement between measured and observed data. This means that the model calibration was done in the right way to access more accurate and real results.

Other simulations (1982, 1995, *and* 2000): Based on the available data, the flow propagation was simulated for the river case study for flood events in 1982, 1995, and 2000, and was compared with the river morphology of 1957 and 2008. The idea behind this kind of simulation was to know how much river channel changes affect flow propagation.

The maximum discharge for all of these years was less than the 2008 flood, and there was also no report about any inundation during these years. Then again, the

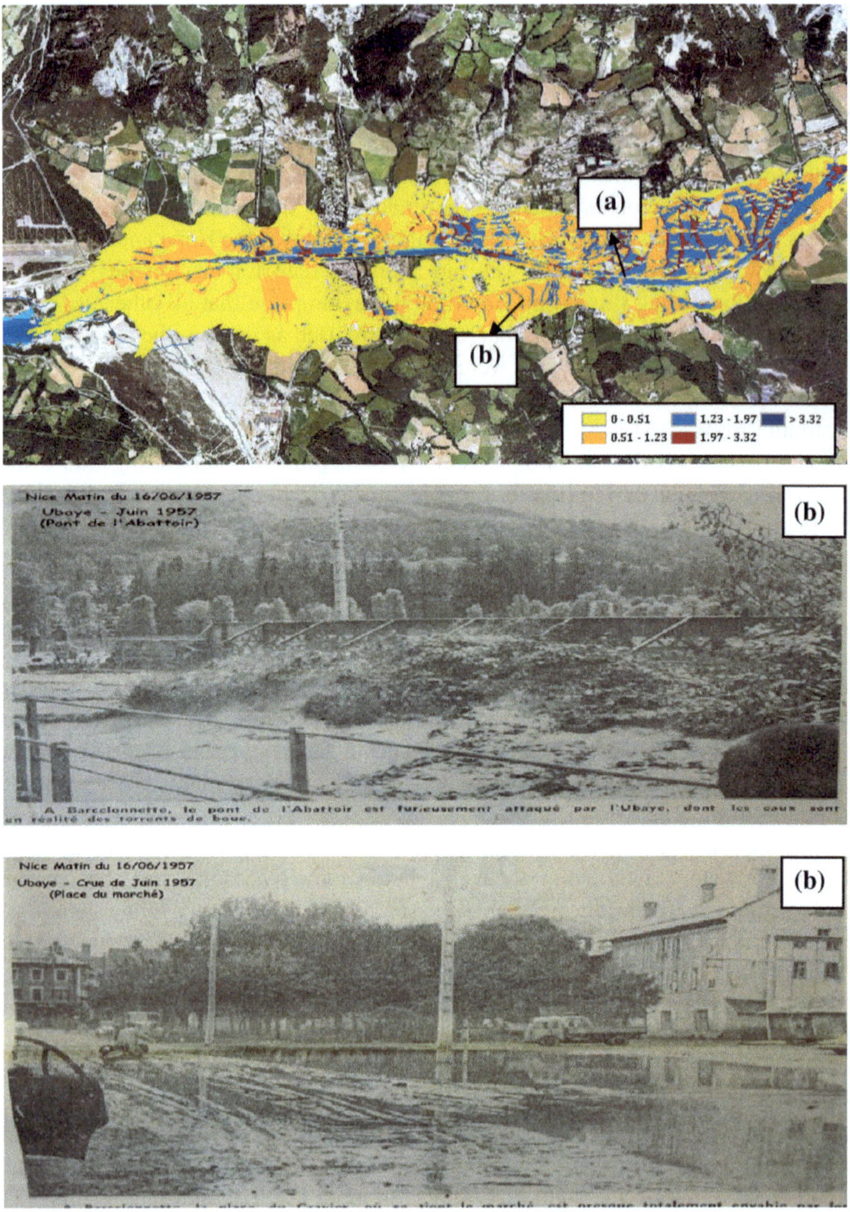

Fig. 6.32 The simulation velocity for flood 1957, validation of inundation with photos. **a** Abattoir bridge; **b** named "Place du marche" or market place (*Source of photos* RTM)

Fig. 6.33 *Left* the Ubaye River flooding in May 2008: a view of a bridge before its destruction; *right* the Ubaye River flooding in May 2008: a view of a bank collapse (*Source* OMIV)

Fig. 6.34 The measurement station, observed height water during flood 2008 (RTM)

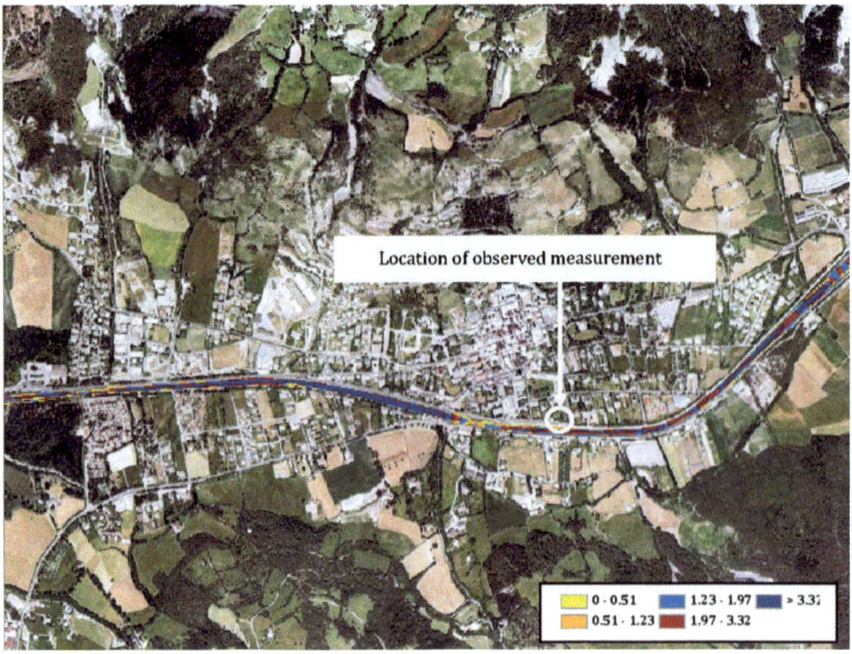

Fig. 6.35 The simulation velocity for flood 2008, validation of based on observed measurements

flow water stayed between the banks. Observed data were not available to calibrate the model for these three simulations; therefore, it seems that these results offer only an idea about the area and river change and show critical places with high flow propagation, velocity, and water depth. Figure 6.36 shows the simulation results for different years from 1957 to 2008.

In general, the following results are obtained from the river change scenario:

(a) Lower near-bank velocity in the environmental part of the river is maybe because of the contribution of vegetation on the banks to higher roughness.
(b) No inundation was observed in natural part of the river. Probably, the vegetation on the banks has a small contribution to overall flow resistance. Restoring vegetation along the river would increase hydraulic resistance.
(c) The right bank was always much more sensitive to the velocity.

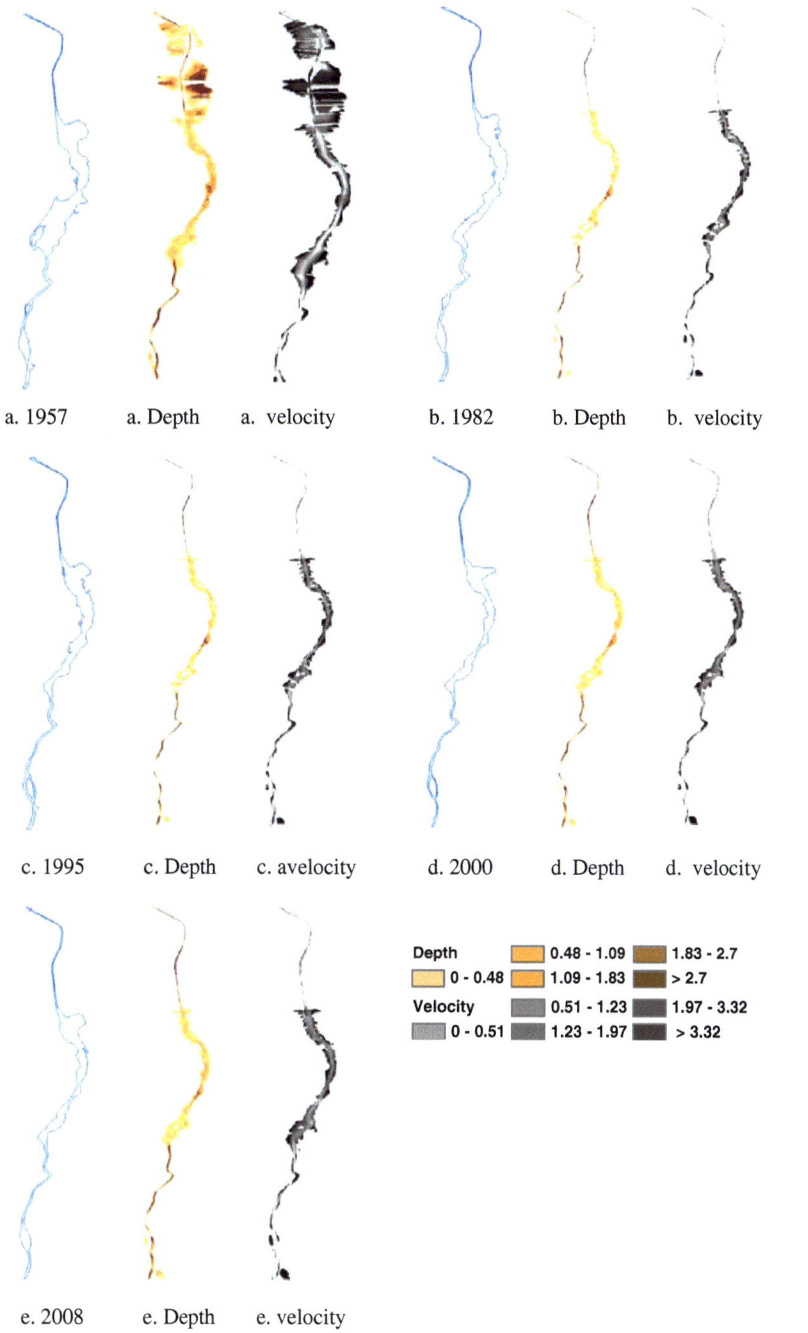

Fig. 6.36 Representation of river change over the time coupled with simulation of velocity (m/s) and depth of water (m). **a** Flood 1957 (area 1.03 km^2). **b** River 1982 (area 0.77 km^2). **c** River 1995 (area 0.74 km^2). **d** River 2000 (area 0.66 km^2). **e** Flood 2008 (area 0.65 km^2)

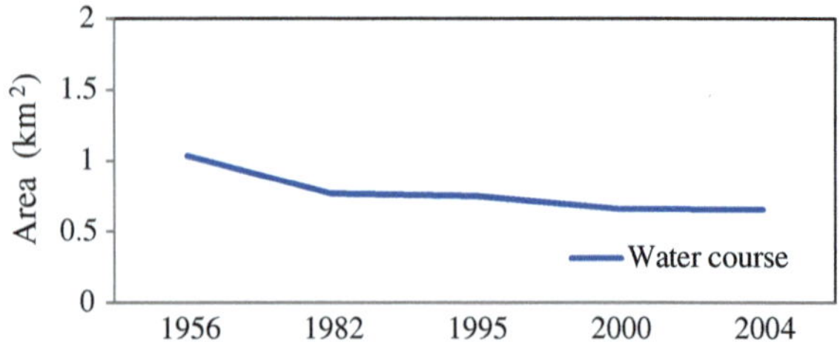

Fig. 6.37 River changes over the time between 1956 and 2004

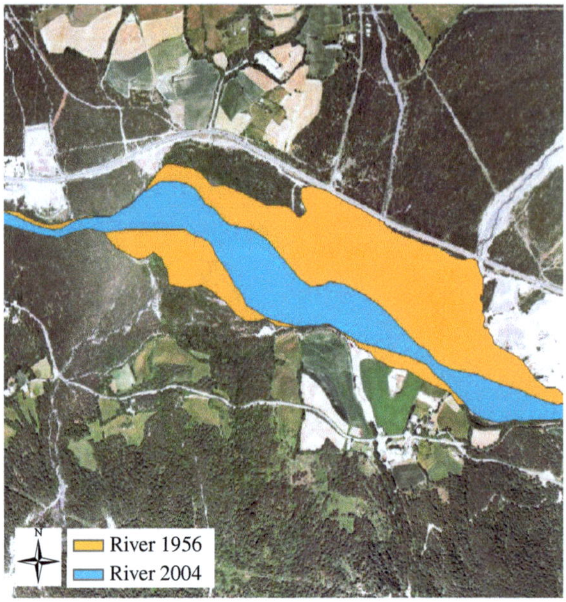

Fig. 6.38 River change comparison between 1956 (*orange color*) and 2004 (*blue color*)

(d) Additionally, these results confirm that constructed dykes in urban areas after flood 1957 could protect the city as well.

River changes over time between 1956 and 2004 show that the river became narrow, especially in middle (Figs. 6.37–6.39). The height of the dyke in the city area was also increased 0.5 m after the 1957 flood to prevent the urban area from any probable flood damages.

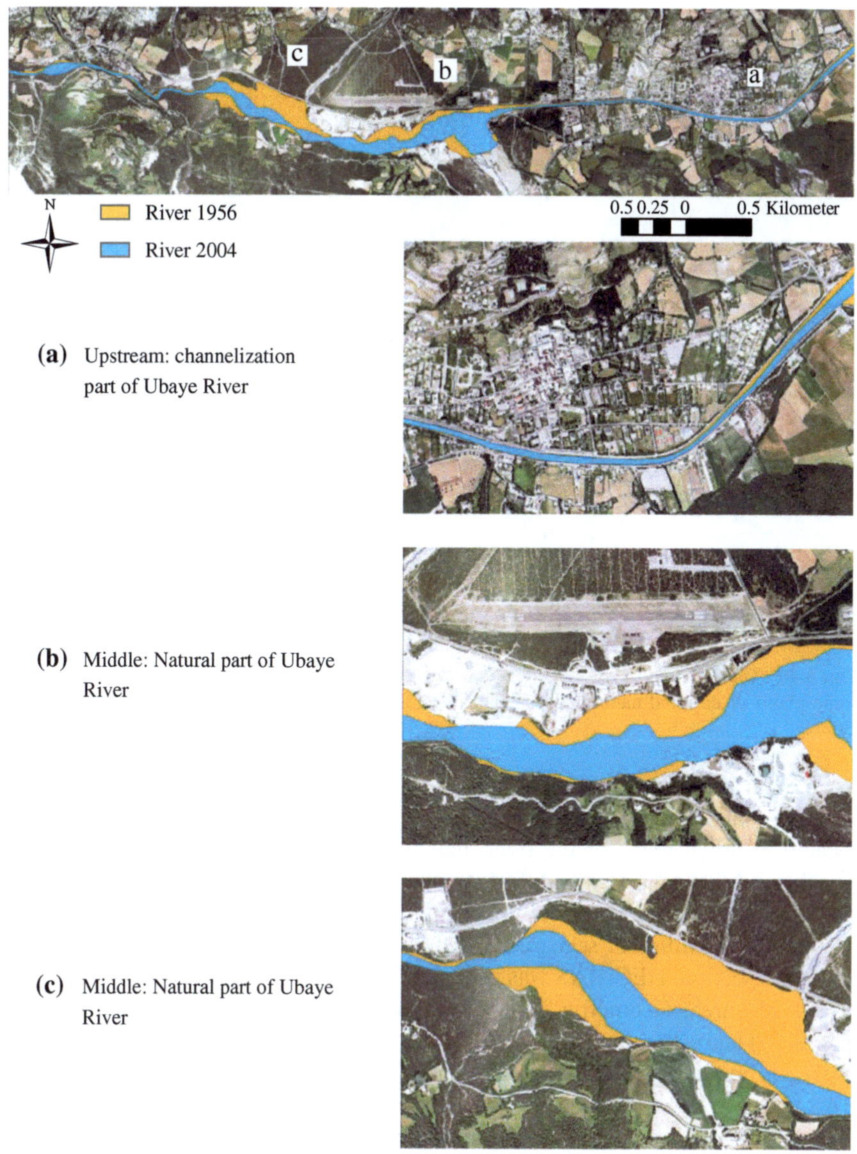

Fig. 6.39 River change; *orange color* represents river 1956 and *blue color* represents river 2004

6.10.3 Dyke Scenario

In case of dyke failure, it is essential to evaluate the possible effects in advance. In this context, simulations of different failure scenarios applying various hydrograph are useful to investigate the river behavior during a flood event. The Ubaye River

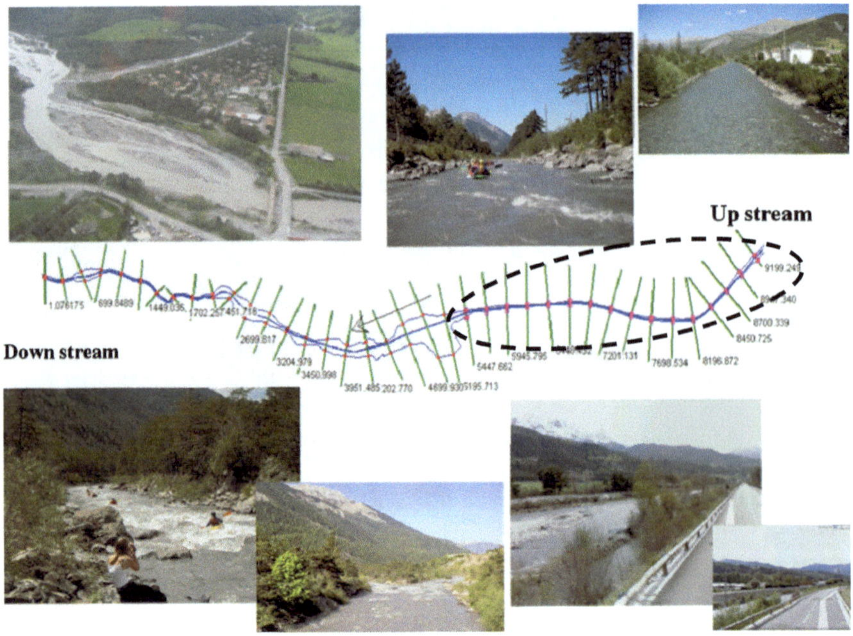

Fig. 6.40 The location of the left dyke and right dyke in urban reach of the Ubaye River with some photo of city and natural part of the river (*Source of photos* Google Earth and RTM)

has a modified and natural river bed. These two parts include a canalized channel in the city and an undisturbed river bed without any infrastructure. The dykes are located in urban reach of the river, having an approximate height of 2 m and a length of 3.5–4 km. Using different criteria, critical flooded areas were determined. By comparing with aerial photos from the region, areas potentially affected by damages in different scenarios could be estimated (Fig. 6.40).

In this section, HEC-RAS was used to model the dyke break scenarios over the case study area. Geometric inputs were achieved with HEC-GeoRAS (ArcGIS extension) and additionally, ArcGIS 9.3.1 was applied for spatial processing of the input and modeling processes, and also to visualize the outputs. Then, the HEC-RAS model was run to investigate the effects of a dyke break in the area. For this purpose, the dyke break was considered on the right and left bank in the city part of the river. The model results show the maximum inundation depth around the dyke break to be about 0.12–1.48 for the 2008 flood event and 0.58–3.20 for the 1957 flood. The following figures show overtopping of the banks as a result of a dyke break in one cross section. The applied analyses include the following scenarios:

Scenario (1): Modeling of the 2008 flood event showed a maximum discharge of 205 m^3/s with the dykes. In case of this scenario (1), normal retention water depth and velocity was shown in the channel–water flow was kept between the banks. The model indicated that the maximum water depth and velocity in the

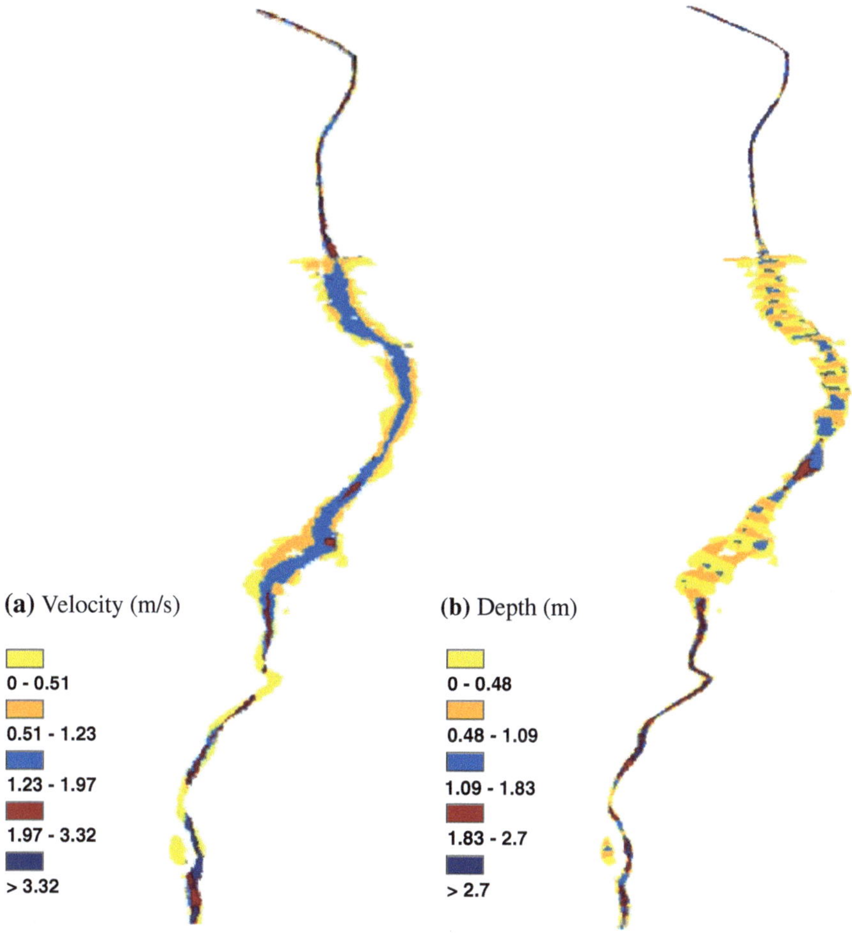

Fig. 6.41 Scenario (1); simulation results of velocity (m/s) and depth of water (m), food 2008; (modeling with considering the dyke)

channel of Ubaye River were approximately 2.7 m and 3 m/s, respectively (Figs. 6.41 and 6.42).

During this event the floodplain was not affected by inundation. The water stayed within the banks, although in some places as shown in Fig. 6.42, the banks became wet and overtopped. It seems that the right bank was also more sensitive to velocity and depth of water comparing to the left bank.

The bridge of "Pont de l'Abattoir" where the measuring station was located had its scale a few centimeters out of the river flow (1.7 m). This result shows the best agreement between measured and observed data. This means that the model calibration was done in the right way to access more accurate and real results.

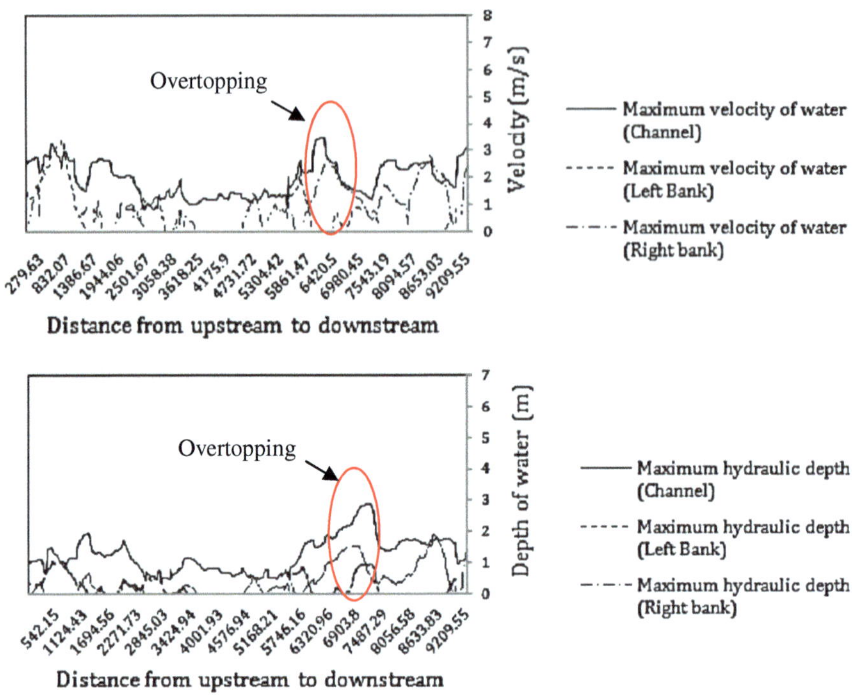

Fig. 6.42 Scenario (1); estimated velocity (m/s) and depth of water (m), food 2008; (modeling with considering the dyke)

In this simulation which was based on flood 2008, if no major flooding occurred, some damages were observed: a bridge had been destroyed, and in some specific areas, river ridges have been scored.

Scenario (2): Modeling of the 2008 flood event showed a maximum discharge of 205 m^3/s without the dykes. Scenario (2) is affected by the channel maintenance and structural dyke integrity. The assumption was that the lack of maintenance produced the dyke break. The results showed that the banks were overtopped by the flood (Figs. 6.43 and 6.44).

By comparing this scenario with previous one, it could be concluded that in case of dyke failure, the capacity of the channel is not enough to keep the water between the banks and in several locations strong overtopping leads to inundation in the floodplain in both sides of the Ubaye River in both city and environmental parts. The velocity and depth of water along the river and across the floodplain is not as stronger as flood 1957 due to lower amount of discharge applied in this scenario.

Figure 6.45 represents sample cross sections located just before and after the dyke break.

Supporting scenario (3): Modeling of the 1957 flood event with a maximum discharge of 480 m^3/s, considering the dyke. This scenario was used as support scenario only for comparing with other scenarios (Fig. 6.46).

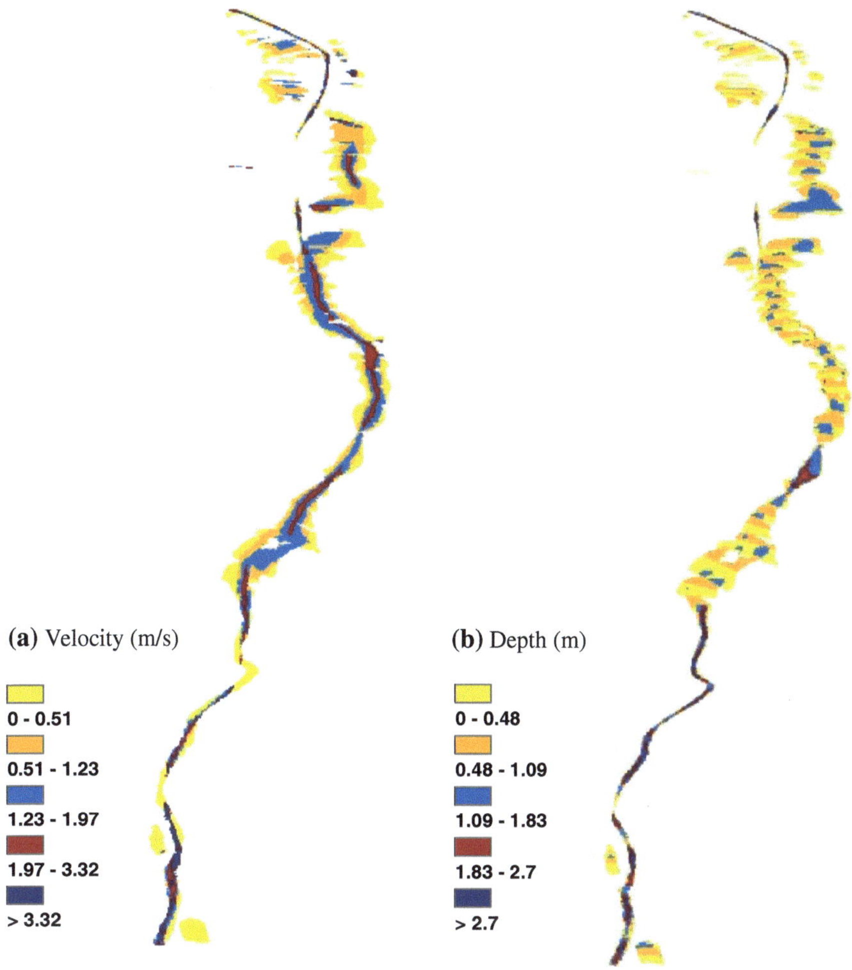

Fig. 6.43 Scenario (2); simulation results of velocity (m/s) and depth of water (m), food 2008; (modeling without considering the dyke)

The 1957 flood event occurred on June 14. Due to a lack of data, there was no possibility to validate the results based on measurements and observed data for water depth and velocity parameters. Therefore, for validation of the simulation results in this case, available photographs obtained from RTM were applied. According to the photographs in Barcelonnette City, the banks of the river were overtopped and inundation occurred in the river floodplain. The velocity (m/s) and water depth (m) for the 1957 flood and modeling without a dyke are shown in Fig. 6.47. The maximum estimated velocity was ~6.32 m/s and the maximum water depth was ~3 m in the channel.

Scenario (4): Modeling of the 1957 flood event with a maximum discharge of 480 m^3/s, assuming the dyke for the river profile of 2008. Figures 6.48 and 6.49

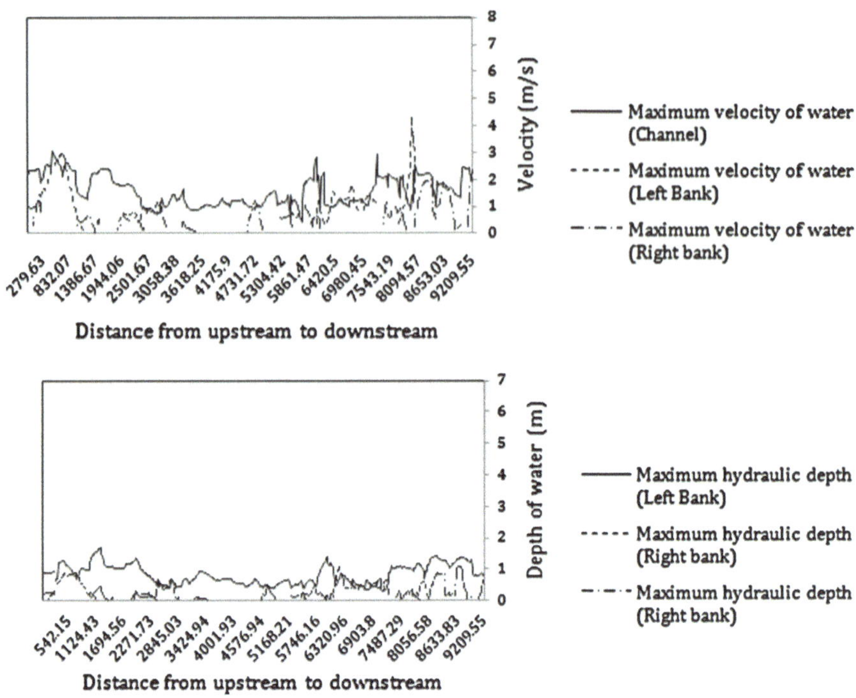

Fig. 6.44 Scenario (2); estimated velocity (m/s) and depth of water (m), flood 2008; (modeling without considering the dyke)

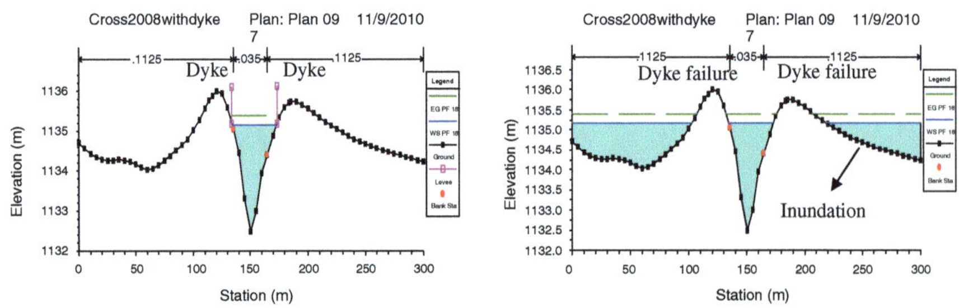

Fig. 6.45 Comparison of flood simulations; *left* before dyke break; *right* after dyke break

show how much adding man-made features to DEM, such as dykes, does effect the flood model simulation and how much the accuracy of the modeling procedure increases (Fig. 6.48).

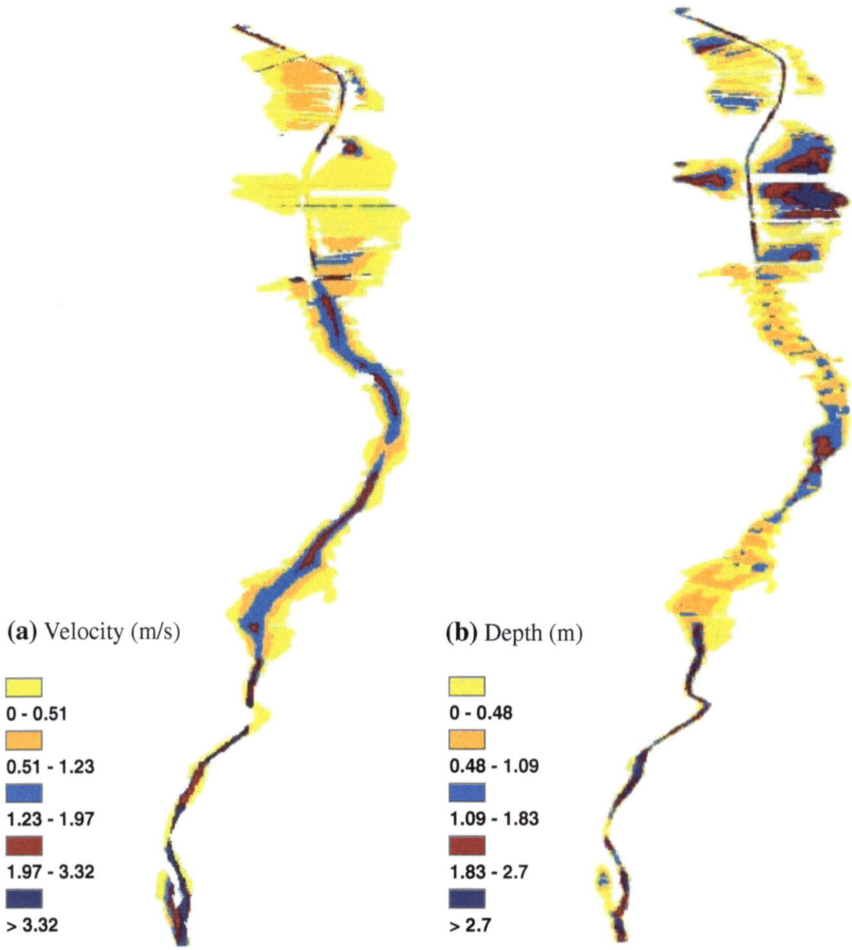

Fig. 6.46 Scenario (3); simulation results of velocity (m/s) and depth of water (m), food 1957; (modeling with considering the dyke)

Comparing the estimated depth and velocity in 1957, the effects for scenario (4) were much more disastrous than in the cases of scenario (1). For this scenario, the same river morphology was assumed as applied for scenario (1). As can be seen in the results obtained from scenario (4) the dykes overtopped on several sections in both banks, especially in the urban sectional area. Critical overtopping occurred

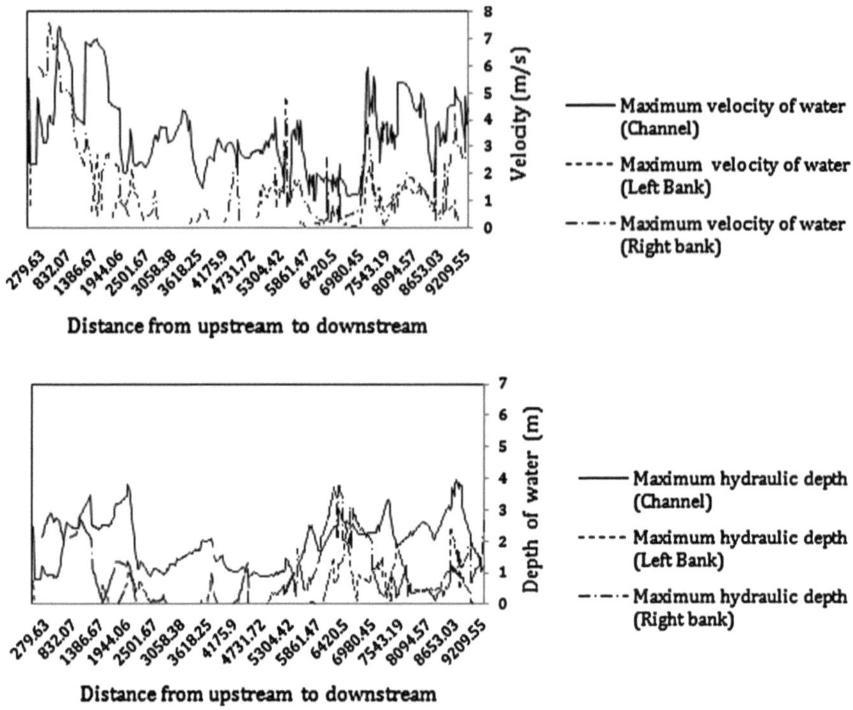

Fig. 6.47 Scenario (3); estimated velocity (m/s) and depth of water (m), flood 1957; (modeling with considering the dyke)

between cross sections 11 and 12 on both left and right banks (~2.5 km from upstream). This shows that at the current morphology of the river and dyke structure, although this hydraulic structure was good strategy to protect the city part of the river in case of severe flood but there is still a concern about overtopping and inundation in case of dyke failure. The simulation results of scenario (4) show that the banks started to overtop between the elevations of 1146.3 and 1147 m where the flow reaches 190 m^3/s (Fig. 6.49).

Figure 6.50 represents the depth of water and velocity parameters for scenario (4). For the last scenario, the modeling of the 1957 flood event had a maximum discharge of 480 m^3/s for the 2008 river profile. The adoption of the dyke break was also considered, and the results of water depth, velocity, and duration are shown in Figs. 6.51–6.53, respectively.

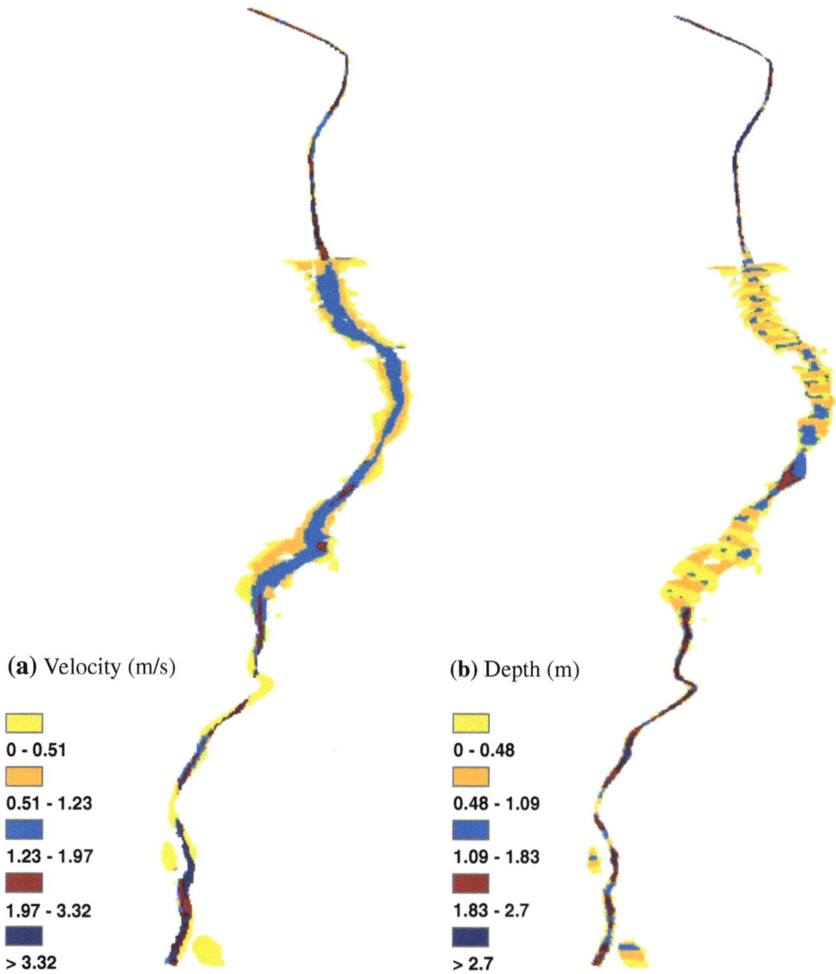

Fig. 6.48 Scenario (4); simulation results of velocity (m/s) and depth of water (m), food 1957 for the profile of 2008; (modeling with considering the dyke)

In the dyke break scenarios, the purpose was to see whether dykes overtop and possibly break from a given flood wave. In the case of a dyke breaking, it calculates the outflow in the hinterland and the reduced water flow volume in the main channel. Figure 6.54a, b, c, and d shows the water volume in the main channel. The left and right banks in the hinterland of the dyke and downstream of the river were also applied to different dyke scenarios.

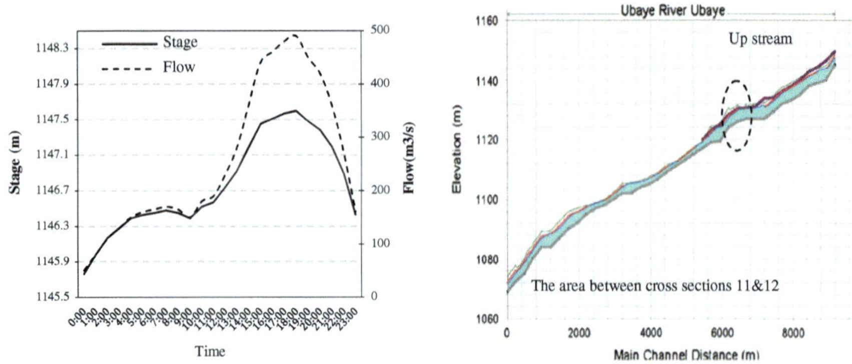

Fig. 6.49 Stage, elevation, flow hydrograph, and time of inundation at cross sections between 11 and 12

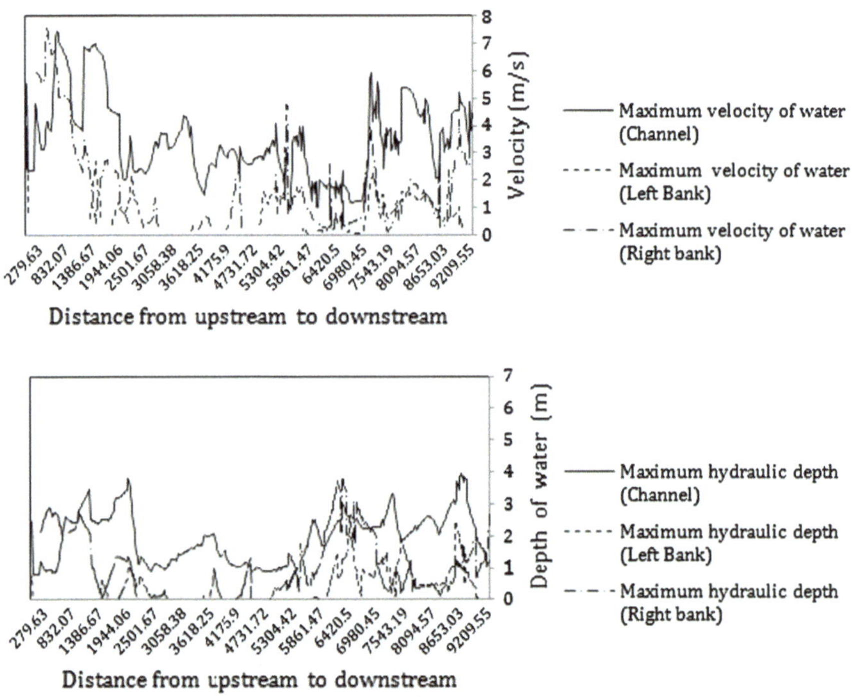

Fig. 6.50 Scenario (4); estimated velocity (m/s) and depth of water (m), food 1957 for the profile of 2008; (modeling with considering the dyke)

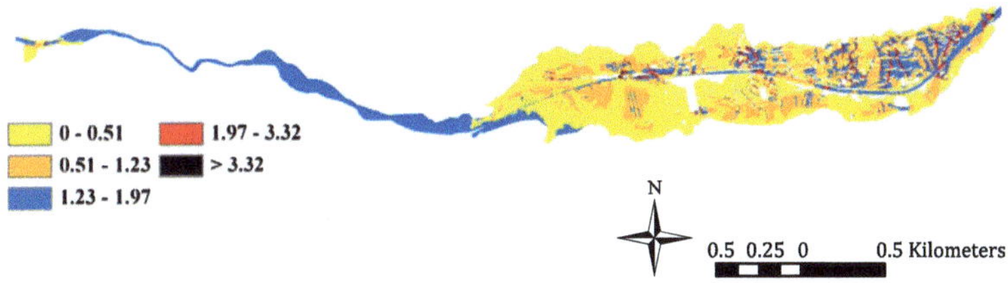

Fig. 6.51 Scenario (4); simulation of velocity (m/s), (food 1957 for the profile of 2008, with the assumption of dyke break)

Fig. 6.52 Scenario (4); simulation of depth of water (m), (food 1957 for the dyke profile of 2008, with the assumption of dyke break)

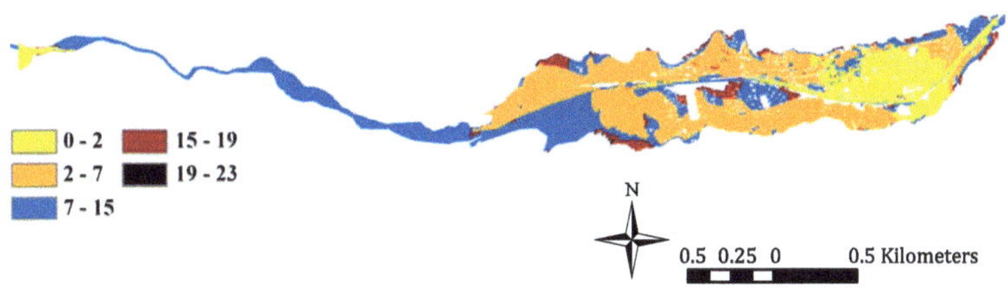

Fig. 6.53 Scenario (4); simulation of duration (h), (food 1957 for the dyke profile of 2008, with the assumption of dyke break)

Figure 6.55 shows the critical depth in Ubaye River profile from upstream to downstream: (a) simulation with maximum discharge of 205 m^3/s and (b) simulation with maximum discharge of 480 m^3/s. As shown in the figures, the critical depth for both models shows almost the same location for upstream and downstream. It means that in these locations, the banks have the capacity to overtop. Figure 6.56 shows the simulation of velocity and water depth in Ubaye River. Dash lines represent simulations with a maximum discharge of 205 m^3/s and solid

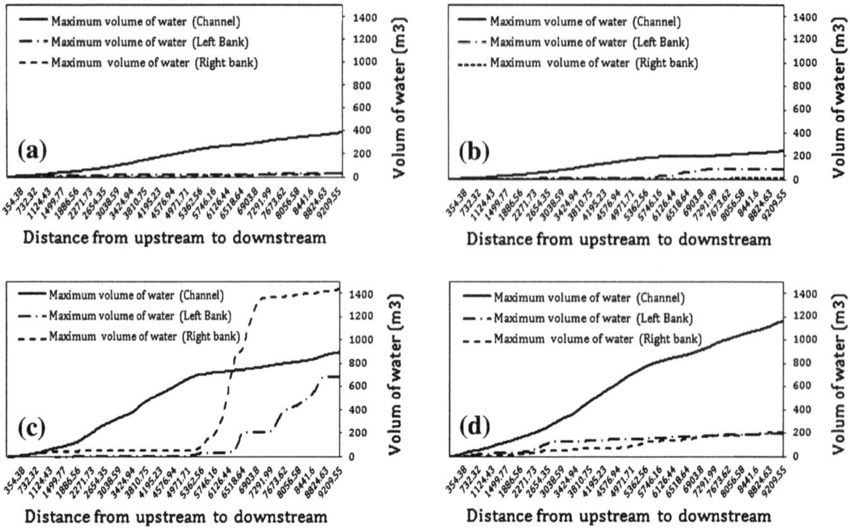

Fig. 6.54 **a** Scenario (1) volume of water. **b** Scenario (2) volume of water. **c** Scenario (3) volume of water. **d** Scenario (4) volume of water

lines represent simulations with a maximum discharge of 480 m^3/s. Upstream is on the left of the graph and downstream is on the right of the graph.

In the results, modeling the maximum flood—inundated—area is delineated in relation to the four cases. Time, elevation, and peak are determined when the flow starts to overtop at one cross section in the upstream and at another cross section in the downstream. Figure 6.57 shows the start of overtopping for upstream and downstream of the Ubaye River in all four scenarios.

Figures 6.58–6.61 represent shear stress (N/m^2) in different scenario simulations which is the force of moving water against the bed of the channel. Gray indicates the critical location to Shear stress.

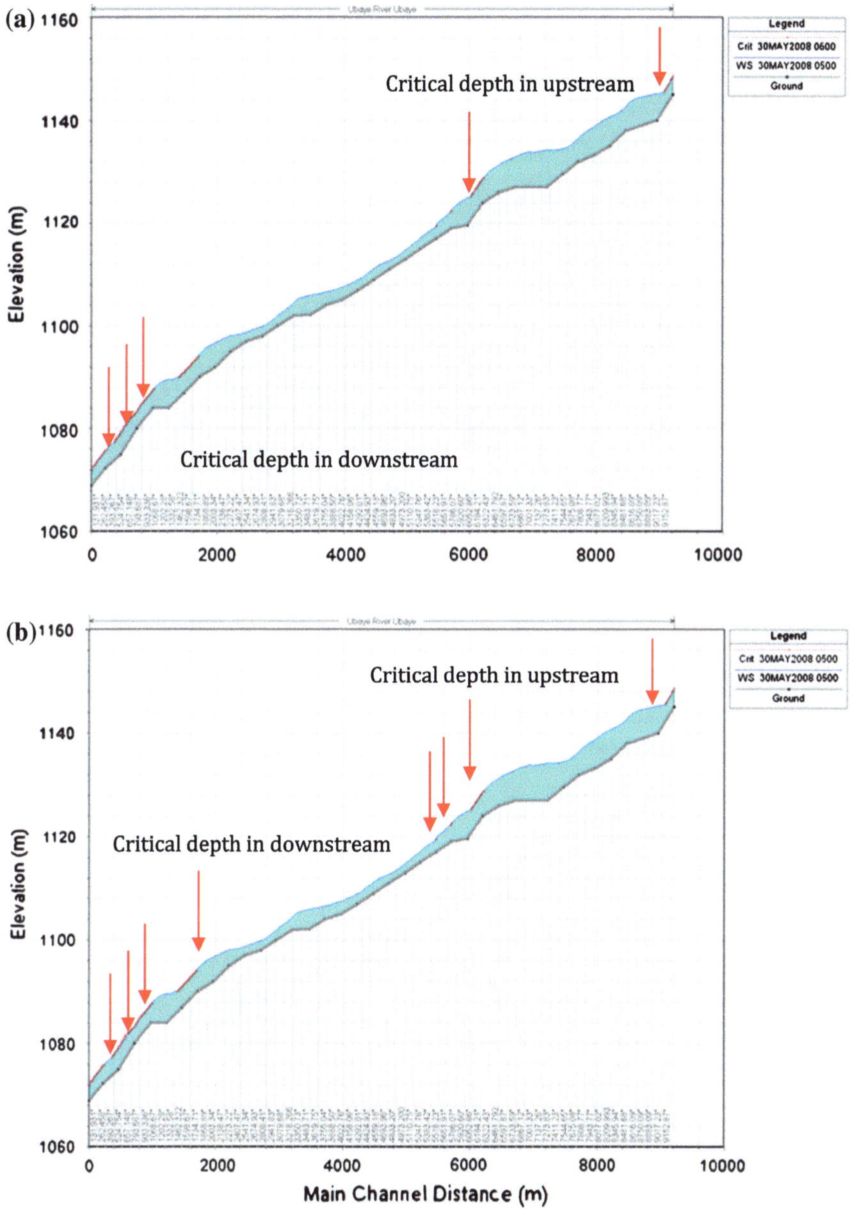

Fig. 6.55 Realizing critical depth in profile of Ubaye River from upstream to downstream. **a** Simulation with maximum discharge of 205 m^3/s. **b** Simulation with maximum discharge of 480 m^3/s

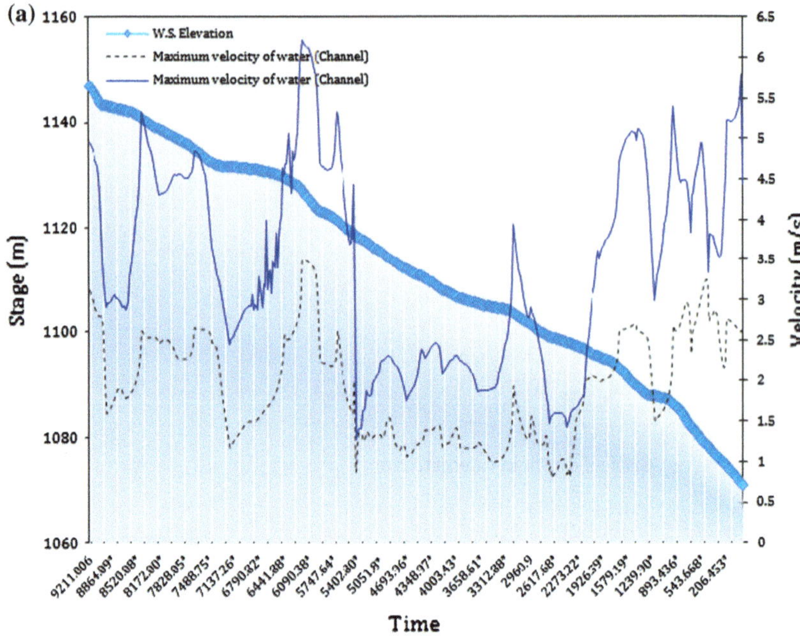

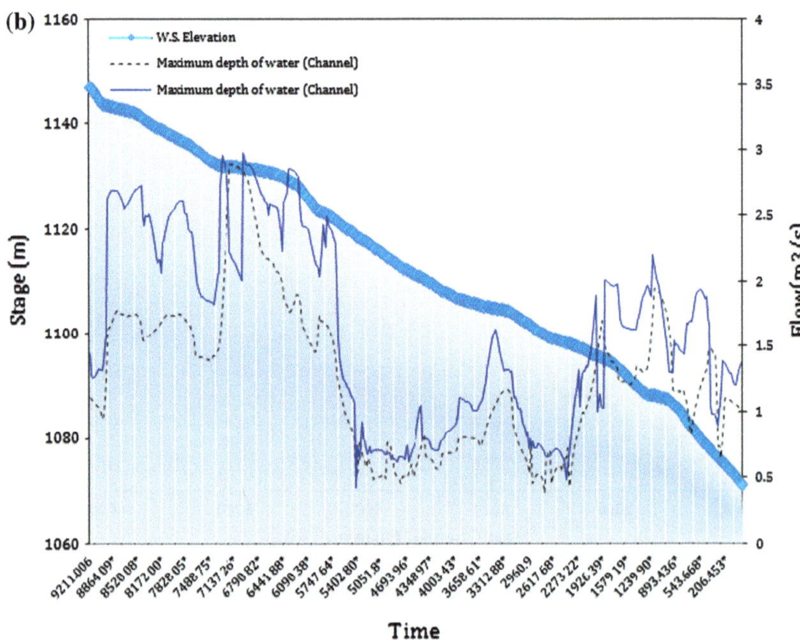

Fig. 6.56 Simulation of **a** velocity; **b** depth of water for Ubaye River; *dash lines* represent simulation with maximum discharge of 205 m^3/s; *solid lines* represent simulation with maximum discharge of 480 m^3/s—upstream link of the graph and downstream right of the graph

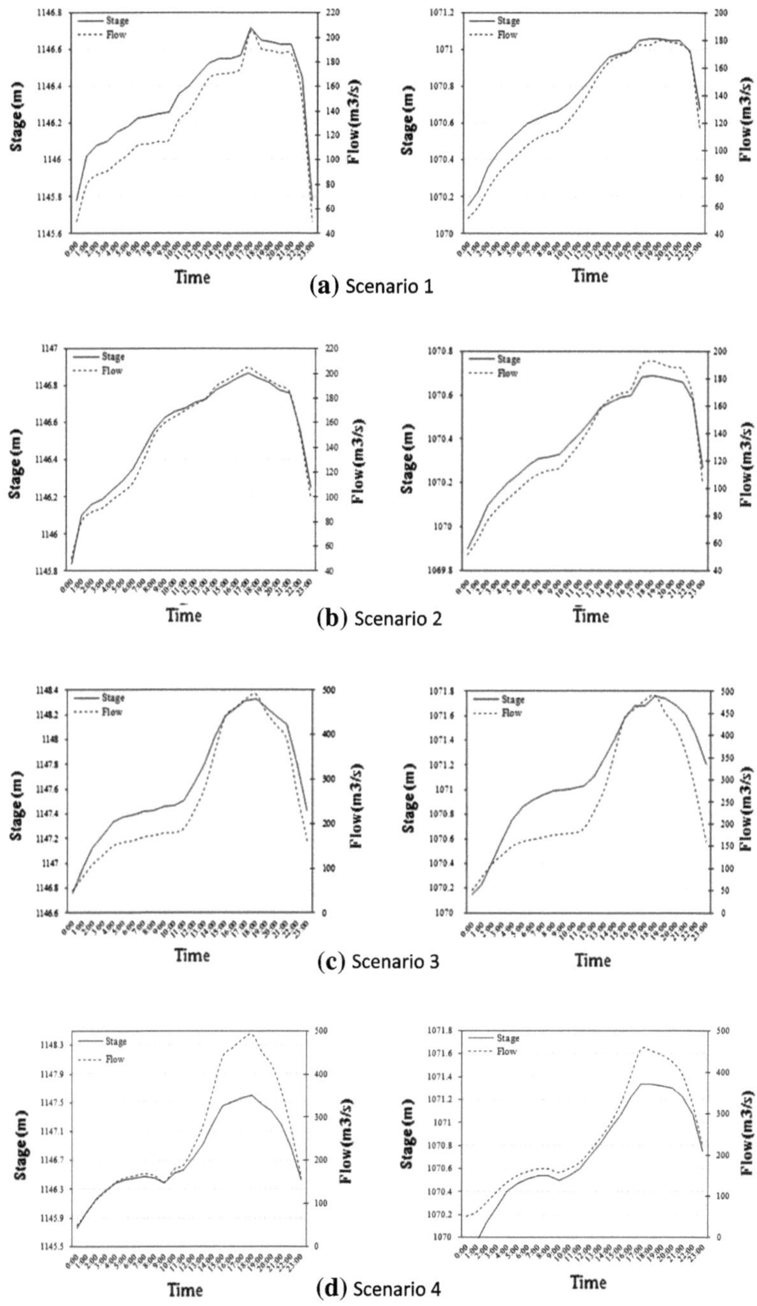

Fig. 6.57 Start to overtopped for upstream and downstream of the Ubaye River for **a** Scenario 1; **b** Scenario 2; **c** Scenario 3; **d** Scenario 4

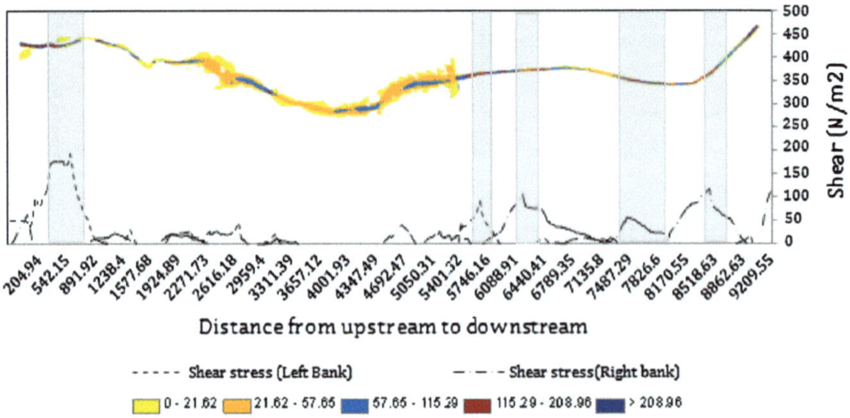

Fig. 6.58 Shear stress for scenario (1); *gray color* represents the critical locations to shear stress

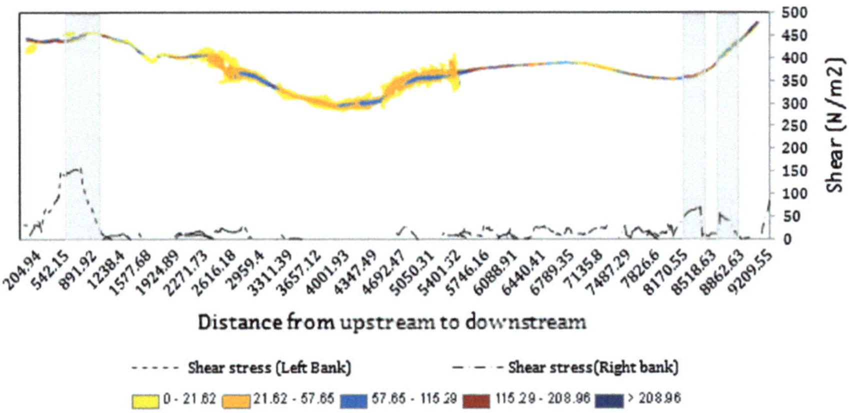

Fig. 6.59 Shear stress for scenario (2); *gray color* represents the critical locations to shear stress

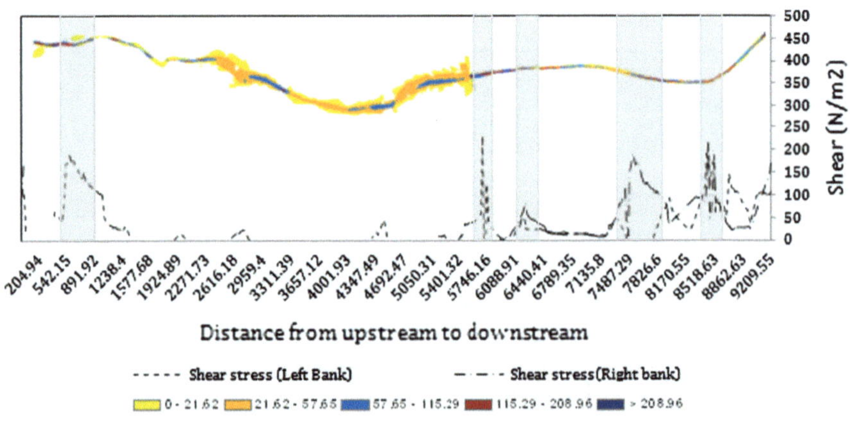

Fig. 6.60 Shear stress for scenario (3); *gray color* represents the critical locations to shear stress

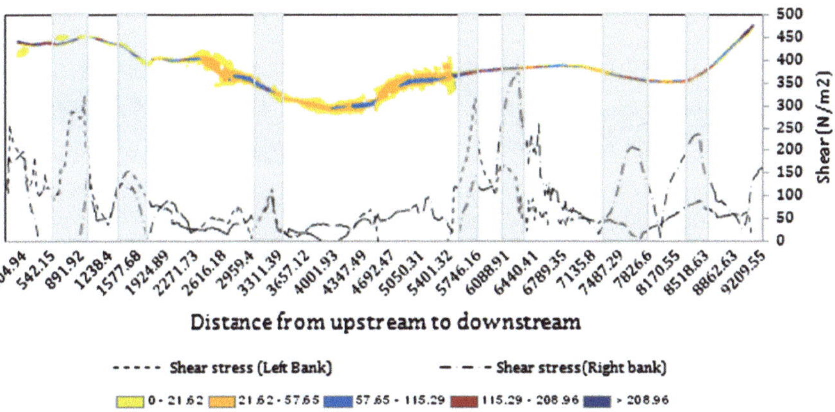

Fig. 6.61 Shear stress for scenario (4); *gray color* represents the critical locations to shear stress

6.10.4 Model Scenario

In the model scenario, the aim was to compare the different results in both HEC-RAS and SOBEK to find the advantages and disadvantages of each model (Table 6.16).

As explained in the last chapters, two different flood modeling packages were used to simulate floods with different DEM resolutions in this study. Figures 6.62 and 6.63 show 1D and 1D–2D simulations in HEC-RAS and SOBEK. Both models are sensitive to the DEM resolution. The main reasons herein are the basic differential equations which deal with the pixel sizes. The models give us different options in flood hazard assessment.

Based on the user's purpose, it has to be determined which model should be applied. If engineers have to decide on dams or house enforcements against flash flows, the SOBEK results are more suitable due to higher accuracy of flow velocity results. However, if only the approximate extent of water is important (e.g., for spatial planning procedures), the HEC-RAS model is sufficient.

The results of flood propagation in both models give better insight into flood characteristics, e.g., where the water moves (inundation) and how fast the water flows (water velocity). These results are very important as basic information for

Table 6.16 Model scenario comparison: HEC-RAS and SOBEK comparison

	HEC-RAS (1D)	SOBEK (1D–2D)
Advantages	1. It simulates better results when water stays within the channel 2. GIS software necessary for data creation, organization, and map publishing 3. Engineering to check and design sediment, velocity, shear stress	1. It has the capacity to model both in 1D and 2D 2. Strong to show inundation area 3. Some results as basis for policy making, decision support and planning such as duration, water level, warning time
Disadvantages	1. Depend on length of cross section to show floodplain interactions	1. Long time for simulation 2. No connection to GIS for data preparation 3. Needs high capacity of RAM

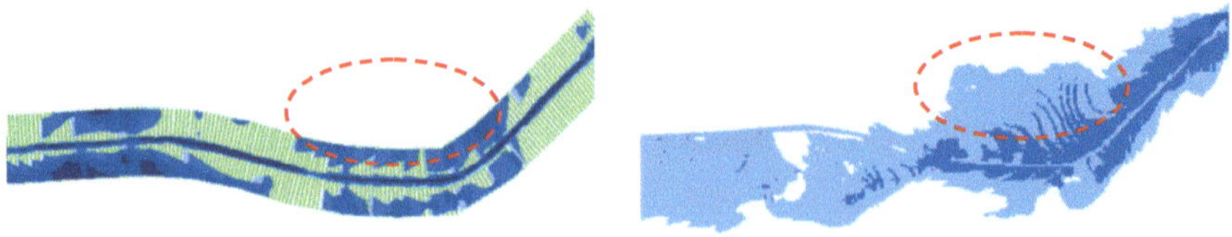

Fig. 6.62 Priority of SOBEK in floodplain simulation

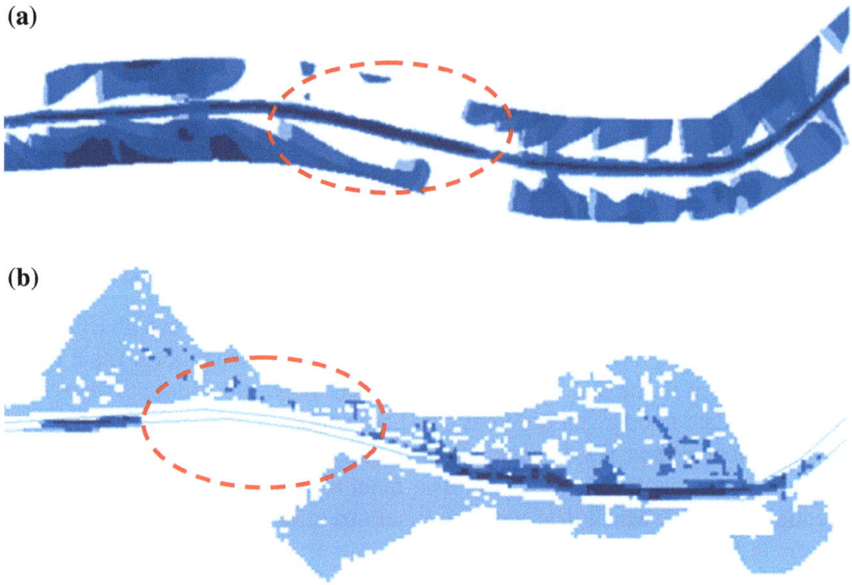

Fig. 6.63 Priority of HEC-RAS in channel simulation; **a** HEC-RAS; **b** SOBEK

policy making and for decisions supporting and planning hydraulic structures. Also, the findings of this section could be more interesting for river hydraulic engineers and researchers.

6.11 Flood Frequency Analysis

6.11.1 Return Period

Flood hazard analysis requires probability or return period statements assigned to a scenario set. Therefore, in this research, synthetic event-based hydrographs based on the extreme value were developed. Table 6.17 shows the results of the Gumbel method to calculate different return periods at both upper and lower levels with 90 % confidence.

Table 6.17 Data preparation for the Gumbel method

Observed discharge *X*-value ranked	Calculated serial number	*X*-value	90 % confidence (lower)	90 % confidence (upper)
18.7	29	18.65	18.65	28.96
30.3	10	30.3	28.84	44.97
34.8	12	34.73	32.43	51.04
42.1	26	42.1	37.78	60.89
42.4	11	42.38	37.97	61.31
50.4	28	50.38	43.21	71.67
54.5	18	54.43	45.71	76.7
59.4	9	59.4	48.59	82.52
61.5	4	61.53	49.81	85.02
65.2	14	65.19	51.91	89.35
70.8	25	70.79	54.93	95.58
76.8	7	76.84	58.16	102.15
80	2	79.97	59.82	105.52
80.5	17	80.41	60.04	105.98
81.5	13	81.45	60.57	107.16
83.2	20	83.13	61.42	108.85
84.7	24	84.7	62.18	110.6
86.4	15	86.32	63.06	112.13
86.5	8	86.48	63.17	112.39
87.3	3	87.31	63.61	113.18
88	6	87.98	63.95	113.98
89.1	27	89.01	64.39	115.06
89.1	31	89.01	64.39	115.06
89.3	19	89.18	64.51	115.34
92.3	21	92.38	66.11	118.43
101.2	5	101.29	70.54	127.48

(continued)

Table 6.17 (continued)

Observed discharge *X*-value ranked	Calculated serial number	*X*-value	90 % confidence (lower)	90 % confidence (upper)
119.4	23	119.31	79.24	145.29
129.3	22	129.18	83.91	154.9
182.9	16	182.84	108.81	205.14
205	30	205.14	118.28	226.12
480	1	494.51	219.12	508.02

X-value ranked	Serial number	Return rank	Return period calculation	90 % confidence (lower)	90 % confidence (upper)
18.7	29	1.03	1.01	1.01	1.07
30.3	10	1.07	1.08	1.07	1.29
34.8	12	1.1	1.13	1.1	1.4
42.1	26	1.14	1.24	1.17	1.63
42.4	11	1.19	1.24	1.17	1.64
50.4	28	1.23	1.39	1.26	1.92
54.5	18	1.28	1.48	1.3	2.07
59.4	9	1.33	1.59	1.36	2.26
61.5	4	1.39	1.64	1.38	2.34
65.2	14	1.45	1.74	1.42	2.49
70.8	25	1.52	1.9	1.49	2.73
76.8	7	1.6	2.1	1.6	3
80	2	1.7	2.2	1.6	3.1
80.5	17	1.8	2.2	1.6	3.1
81.5	13	1.9	2.2	1.6	3.2
83.2	20	2	2.3	1.6	3.3
84.7	24	2.1	2.3	1.7	3.3
86.4	15	2.3	2.4	1.7	3.4
86.5	8	2.5	2.4	1.7	3.4

(continued)

Table 6.17 (continued)

X-value ranked	Serial number	Return rank	Return period calculation	90 % confidence (lower)	90 % confidence (upper)
87.3	3	2.7	2.4	1.7	3.5
88	6	2.9	2.4	1.7	3.5
89.1	27	3.2	2.5	1.7	3.5
89.1	31	3.6	2.5	1.7	3.5
89.3	19	4	2.5	1.7	3.6
92.3	21	4.6	2.6	1.8	3.7
101.2	5	5.3	2.9	1.9	4.1
119.4	23	6.4	3.7	2.2	5.1
129.3	22	8	4.2	2.3	5.7
182.9	16	10.7	8.4	3.4	10.5
205	30	16.0	10.5	3.9	12.8
480	1	32	61.2	10.8	66.3

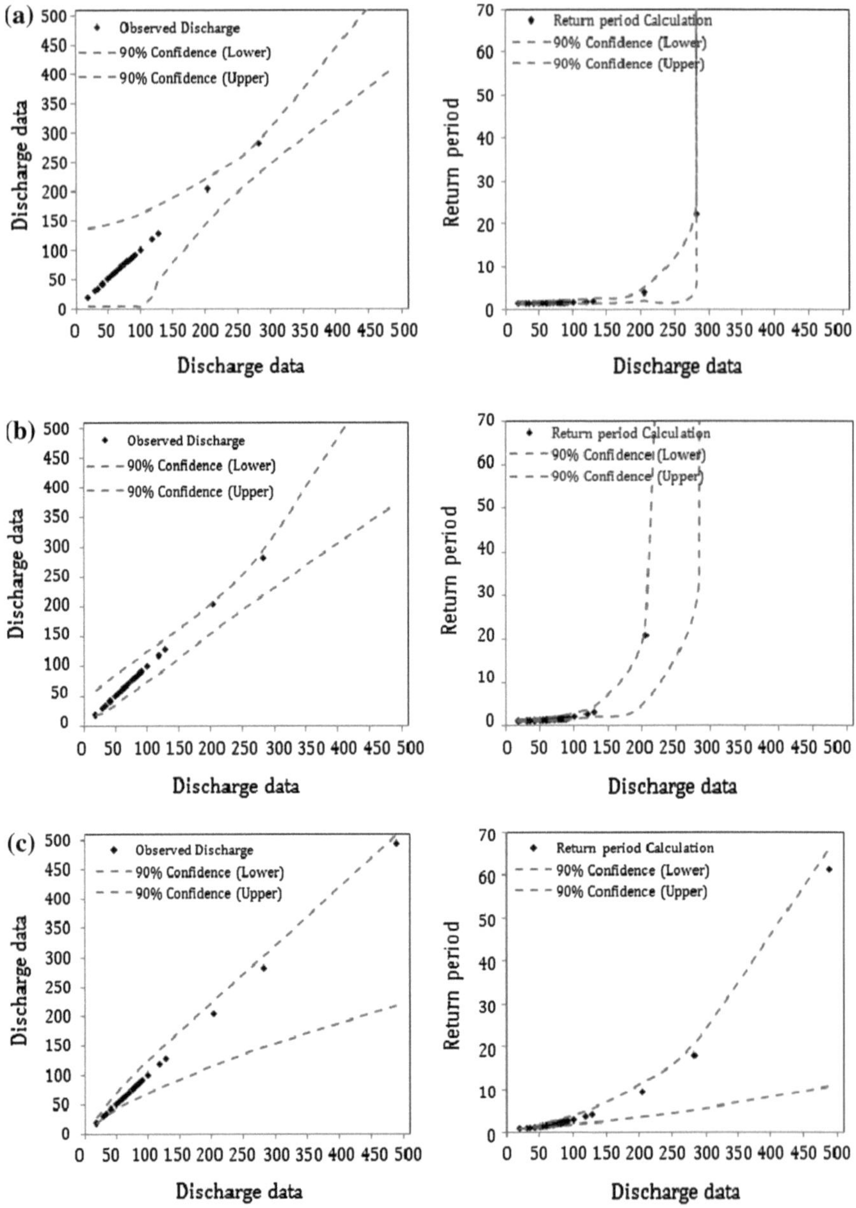

Fig. 6.64 The results of calculated discharge and return period analysis with 90 % confidence for different methods; **a** square normal; **b** normal; **c** Gumbel

Figure 6.64a, b, and c shows the calculated discharge and return period analysis with 90 % confidence in lower and upper parts for the different selected methods. As shown in Fig. 6.61a, b, and c, the best fitted results belong to Gumbel methods.

The results of statistical methods for frequency analysis of the Ubaye River in Barcelonnette show that the 1957 flood event may, on average, be expected annually every 11 years with a 90 % lower confidence level and every 66 and 100 years with a 90 % upper confidence level.

6.11.2 Gumbel Extreme Value Distribution

The objective of these distributions was to build the relation between the probability of the occurrence of a certain event, return period, and its magnitude. To this purpose the Gumbel Method was applied in this research. Table 6.18 shows the calculated discharge, return period analysis and probability analysis with 90 % confidence in lower and upper levels.

Figure 6.65 represents the hydrograph for different return periods based on the Gumbel method.

Figure 6.66 shows a simulated flood hazard for a different return period of 66 years (blue) with 0.015 probability of occurrence, and of 100 years (red) with 0.01 probability of occurrence as well. Orange represents a 100-year return period when implementing a dyke break. In both return periods of 66 and 100 years there is strong overtopping on the right and left banks, causing higher risk to the banks. The results of this overtopping are the banks getting wet and a discontinuous flow in the floodplain as inundations. By assuming a dyke break, the inundation is clearly obvious, and it posed a higher risk to nearby elements (e.g., houses and farm lands). Flood hazard maps should cover the geographical areas which could be flooded according to different scenarios:

- Floods with a low probability or extreme event scenarios
- Flood with a medium probability (likely return period of 66 years)
- Floods with a high probability

One of the aims of this section was to investigate how flood frequency distributions along the river reaches are influenced by the adoption of dyke breaks. The simplest method for the flood frequency analysis in terms of flood hazard assessment was transferring of hydrograph of different return periods at Barcelonnette gauge.

In some cases, it is applied to the situation with and without dykes, thus giving a rough idea of the dykes' flood defense effects. In flat lowland areas located at the intersection of DEM and water levels at the gage, unrealistic inundation extends

Table 6.18 Return period and its probability based on flood 1957 (maximum discharge of 480 m^3/s)

	Maximum discharge	Return period	Probability
90 % confidence (lower)	210	11	0.09
90 % confidence (upper)	507	66	0.015
Over topped	600	100	0.01

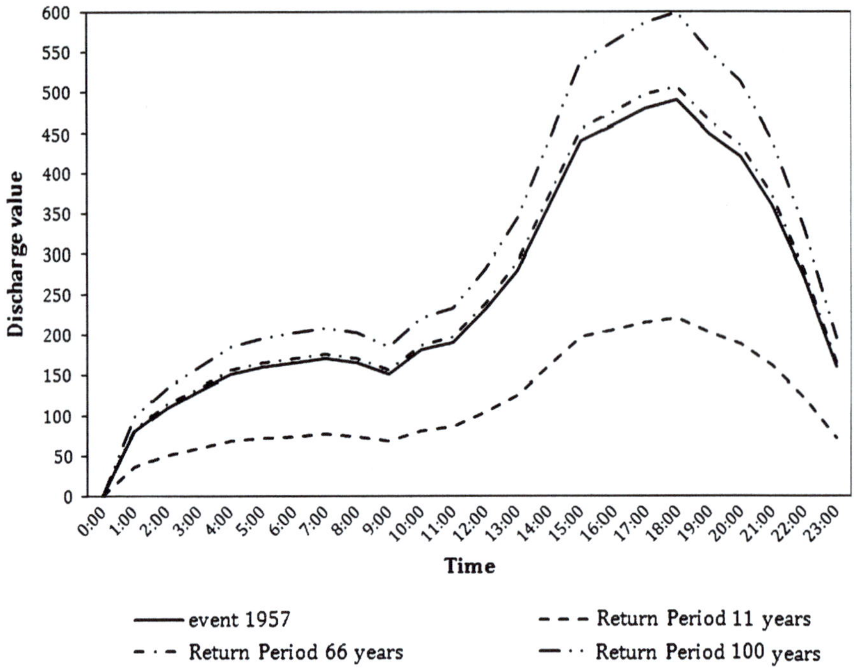

Fig. 6.65 Calculated discharge and return period analysis with 90 % confidence level

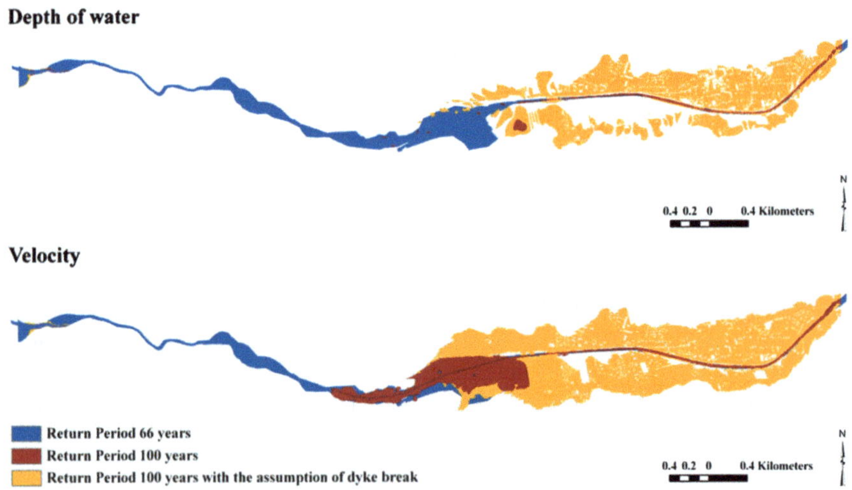

Fig. 6.66 Flood simulation for different return periods

might be produced, because the inundation area may be limited by the water volume available for flooding. More sophisticated methods use 1D or 2D hydrodynamic models to simulate the flooded area associated with a certain discharge value. Such approaches can include the effects of dyke break.

However, since the T-year discharge for certain river sections is taken from the flood frequency analysis, the effects of upstream dyke breaks does not propagate, and they do not affect the flood frequency analysis at downstream. The graphs of dynamic simulations indicate the temporal variability of the flow velocities in the main channel and inundated area. Simulated velocity values were categorized between 0 and 0.51 to >3.32 for return periods of 11, 66, and 100 years.

The local risk protection authority—Service de Restauration des Terrains de Montage—Mountain and Restoration Service (RTM) and Plan d'Exposition aux Risques (PER), in combination with the Municipality, had prepared maps for risk assessment based on multiple natural hazards for the case study region. They were collected during the field from the local municipality and gave insight into the type of protection work that has been carried out based on these maps (PER; RTM; [1]). RTM is one public administration managing the natural risks and has been created in 1860 after some huge torrential events in the French Alps.

Repairing and maintaining the dykes and embankments, reforestation of the slope, maintenance of upstream check dams and plants for building new bridges are the major protection activities that are based on these risk maps. The maps were based on multi-hazard criteria such as landslides, debris flow, inundation, and mudslides. Communication with RTM officials revealed that the basis of flood risk assessment include: the capacity for water to pass through the bridge on the Ubaye River, the mass movement caused by the torrent in La-vallette, houses at the upper slopes at risk, and experts views. The area was divided into low, medium, and high risk zones based on the intensity of probable risks in the area (PER; RTM; [1]).

The methodology for delineating flood hazard zones as a result of torrential activities and inundations identified by Plan d'Exposition aux Risques (PER) and Restauration des Terrains de Montage—Mointain alnd Restoration Service (RTM) during 1985–1995 is presented in Fig. 6.67a. This map is compared with simulated flood maps with 66 and 100-year return period, with and without a dyke (Fig. 6.67b).

Simulation results of velocity and water depth parameters for a 100-year return period with the assumption of a dyke break were also compared with the flood hazard zone, provided by Prévention des Risques Naturels (PPR) since 1995 onward (Figs. 6.68 and 6.69).

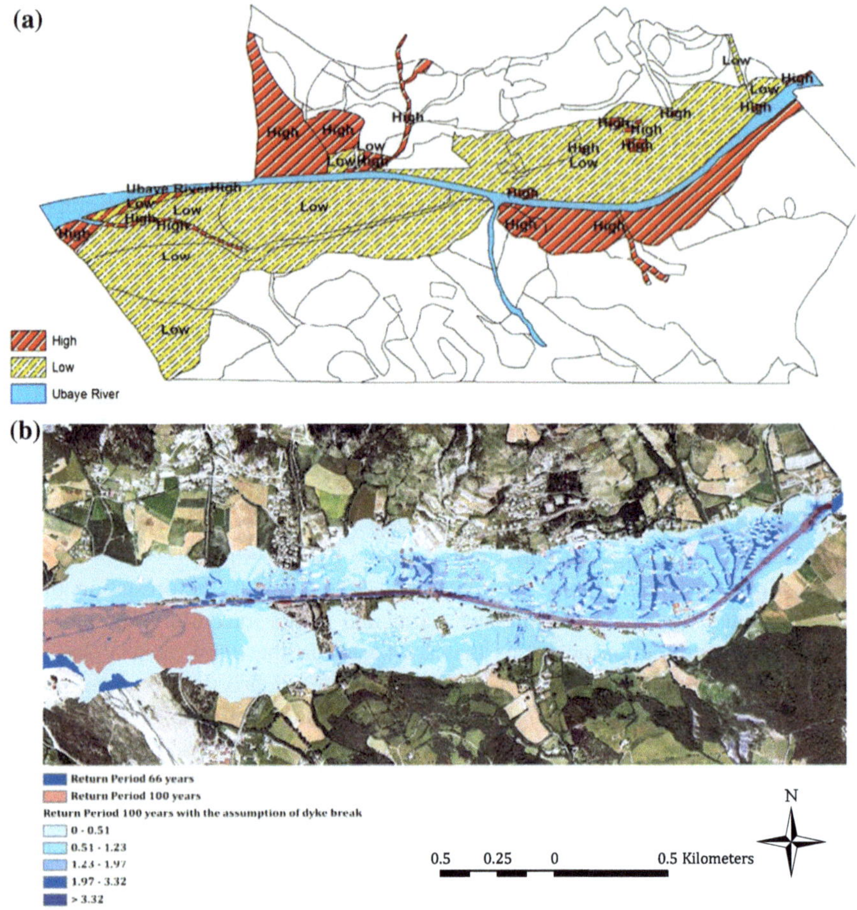

Fig. 6.67 Comparison between **a** risk map for torrential an inundation provided by RTM during 1985–1995 for Barcelonnette city; and **b** simulation results for different return periods of 66, 100 years with dyke and 100 years with the assumption of dyke break

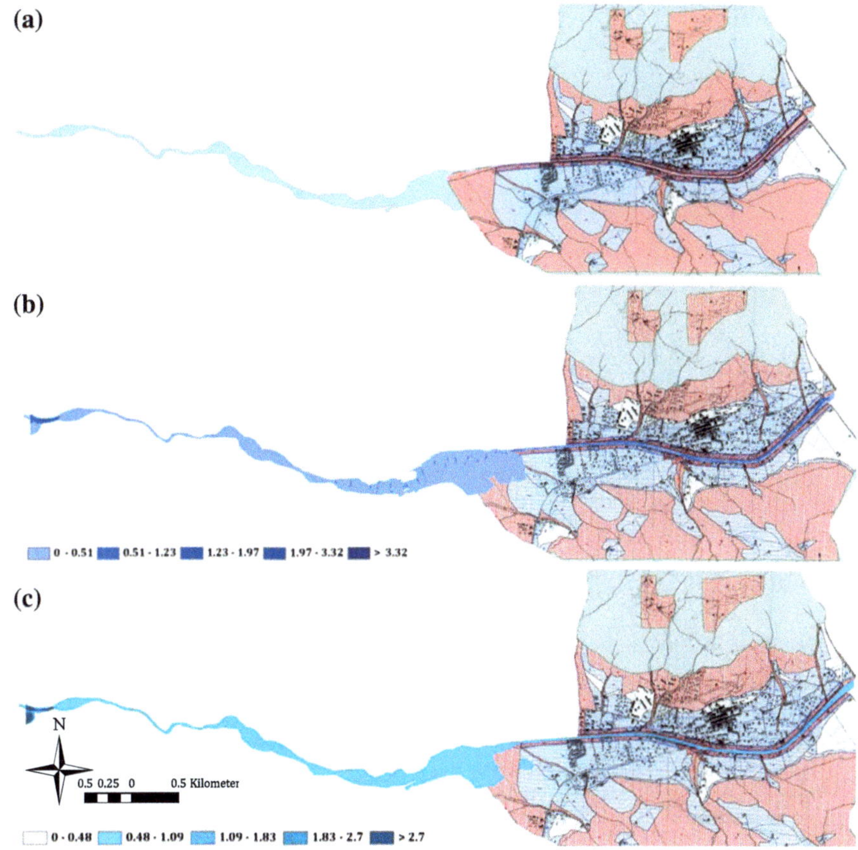

Fig. 6.68 **a** Multi hazard zones provided by PPR since 1995 onwards (the *red* and *blue colors* correspond to channel and inundation zones); **b** flood hazard map (velocity) for 100 year return period; **c** flood hazard map (depth of water) for 100 year return period

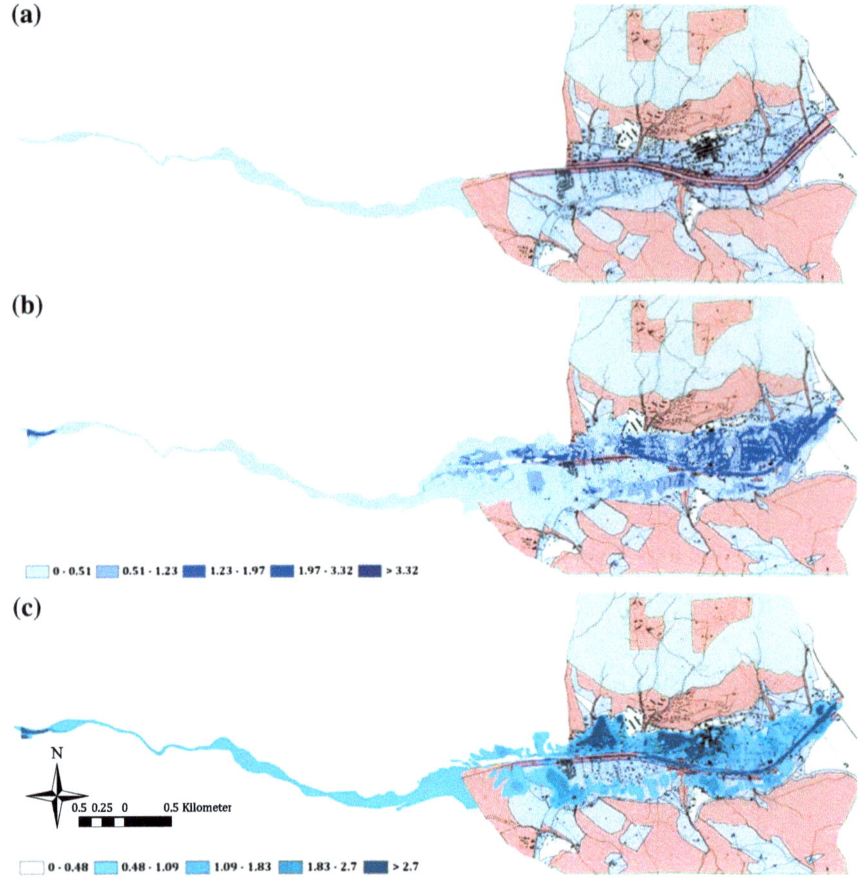

Fig. 6.69 **a** Multi hazard zones provided by PPR since 1995 onwards (the *red* and *blue* colors correspond to channel and inundation zones); **b** flood hazard map (velocity) for 100 years return period with the assumption of dyke break; **c** flood hazard map (depth of water) for 100 years return period with the assumption of dyke break

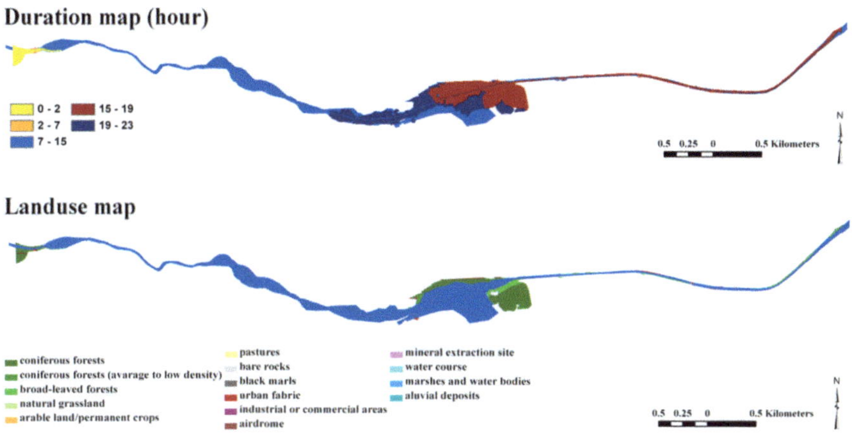

Fig. 6.70 Flood simulation for 100 years return period and affected land use map

The results of velocity and water depth simulations for 100-year return periods show that the reconstruction of dykes after the 1957 flood event was a good strategy to protect the area from such a severe flood. In several locations in the channel on the left and right side of the river, strong overtopping was realized, matching with the red zones on the PPR map. The idea behind the simulation represented in Fig. 6.66 was to consider the dyke break implementation in a 100-year return period.

From Fig. 6.69, it can be observed that the simulated results with the adoption of dyke breaks that were shown on the maps for a 100-year return period corresponded with the zone limits delineated by hazard maps by PPR. High zones in the channel were especially in agreement with the PPR map.

The results of warning time maps for a return period of 100 years are shown in Fig. 6.70. This is the minimum time period that people have to be relocated, that businesses and industries are closed and that transportation in and through the area might be impossible or hindered. It is an important parameter for assessing the economic and social impacts of the flood on the people's life and work conditions in the area. It is also an important parameter for estimating agricultural damage, because many crops, like fruit bearing trees and vineyards, can withstand inundation of their stems for a short time (usually some days), but when the length of time becomes too long the roots will starve from oxygen depletion and the trees will die.

Table 6.19 shows the affected area and type of the land use which were affected by a flood of a 100-year return period. The areas which were most affected were coniferous forests, with an area of about 0.18 km^2 that is located in middle part of the Ubaye River.

With the assumption of a dyke break (Fig. 6.71) for a 100-year return period, the affected area is about 2.19 km^2, of which 49 % belongs to urban area, 21 % to arable land, and 30 % to coniferous forests (Table 6.20).

Table 6.19 The area affected by flood for return period of 100 years

Type	Area (km^2)	Percent
Coniferous forests	0.18	94
Arable land	0.00	1
Urban fabric	0.01	5

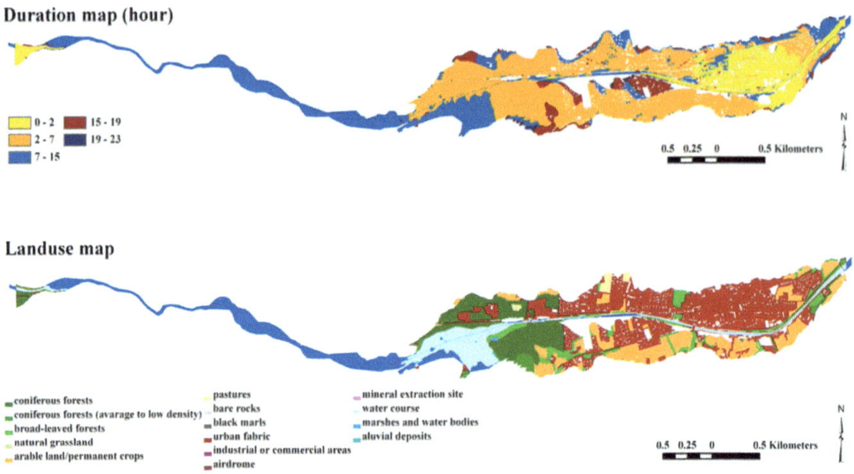

Fig. 6.71 Flood simulation for 100 years return period with the assumption of dyke break and affected land use map

Table 6.20 The area affected by flood for a 100-year return period

Type	Area (km^2)	Percent
Coniferous forests	0.66	30
Arable land	0.46	21
Urban fabric	1.07	49

6.12 Trend Detection in Hydrometeorological Variables

The study area has a dry and mountainous Mediterranean climate (OMIV) with strong inter-annual rainfall variability. Observed data at the Barcelonnette station over the period from 1961 to 2009 show minimum annual precipitation of 358.2 mm, a maximum of 1074.8 mm and an average of 660.30 mm. There are five stations close to the Bracelonnette station (X: 945220.243; Y: 241478.237; Altitude: 1155 m), called Sain-Paul (X: 957569.788; Y: 264628.877; Altitude: 1903 m), Condamine–Chatelard (X: 950635.645; Y: 249358.069; Altitude: 1325 m), Jausiers (X: 9522314.44; Y: 243388.73; Altitude: 1510 m), Uvernet-Fours (X: 947298.32; Y:

233480.074; Altitude: 1660 m), and Allos-Nord (X: 942976.242; Y: 223000.804; Altitude: 1370 m). These stations are considered in climate change detection.

6.12.1 Detection of Randomness

This section presents the results of randomness tests for selected variables. The test of randomness is done by drawing the autocorrelation functions (ACFs) for monthly data including precipitation, maximum discharge and maximum temperature for the Barcelonnette station, and precipitation for other five stations. The results of randomness for all stations showed that all variables are random and autocorrelation coefficients are within 95 % confidence levels, which provide evidence of randomness of precipitation, discharge, and temperature. As a sample, the ACFs of precipitation discharge and temperature for the Barcelonnette station are showed in Fig. 6.72.

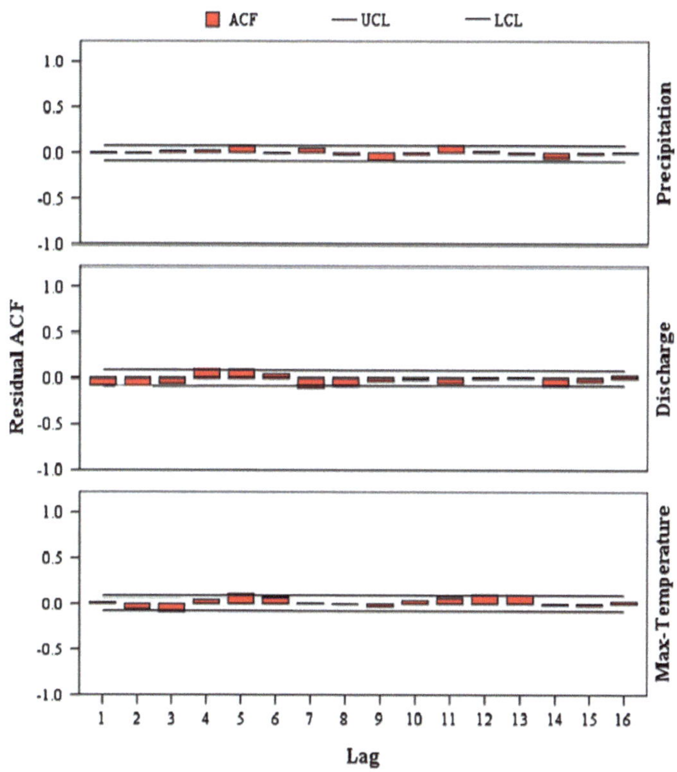

Fig. 6.72 Detection of randomness for precipitation, discharge, and temperature time series

Table 6.21 Statistics of homogeneity test (z-value) for selected variables

Runs test

Parameters	Station	z	95 % confidence interval (P value)
Precipitation	Barcelonnette	*−2.063*	*.039*
Discharge	Barcelonnette	*−11.392*	*.000*
Max-temperature	Barcelonnette	*−16.014*	*.000*
Min-temperature	Barcelonnette	*−16.180*	*.000*
Precipitation	Jausiers	−1.24	.216
Precipitation	Uvernet	−0.49	.620
Precipitation	Condamine	−0.39	.696
Precipitation	Allos	*−1.66*	*.095*
Precipitation	SainPaul	*−1.78*	*.075*
Median			

6.12.2 Detection of Homogeneity

Statistical analysis was done on the available monthly data of precipitation, maximum discharge (these data are maximum values of discharges which occurred at the Barcelonnette station based on recorded hourly discharge), and maximum temperature during the period from 1961 to 2009. Due to a lack of high temporal resolution precipitation data, only the available monthly precipitation series was used in this study. Z-score is known as a standardized value. To obtain z-scores for a variable for each case, subtract the variable's mean value and divide it by the standard deviation. Z-scores are useful for finding outliers and for comparing values of variables that are measured on different scales. The Z value of the homogeneity test was analyzed in the SPSS package. The z-score column is compared to a standard normal table. Thus, at a 5 % significance level, a Z-score with an absolute value greater than 1.96 indicates nonhomogeneous. The significant values are presented in bold italic format. In the period between 1961 and 2009, three stations, Jausiers, Uvernet, and Condamine, represent nonsignificant Z-values for precipitation. On the other hand, precipitation variables in Sain-Paul and Allos-Nord represent significant Z-values. At the Barcelonnette station, the Z-value is also significant with 95 % confidence in the lower and upper parts for precipitation, discharge, and maximum/minimum temperature parameters. This may imply the existence of abrupt climate change in these stations (Table 6.21).

At two other stations close to Barcelonnette, Allos-Nord and Saint Paul, there is an abrupt point in precipitation where the descriptive statistics of the variables (such as mean and variance) begin to change suddenly. To find the time of this change, a sequential Mann–Kendall test was applied.

6.12.3 Detection of Trend: Mann–Kendall Test

A nonparametric Mann–Kendall test was used to determine the trends of the hydroclimate variables of precipitation, discharge, and temperature. This section shows the results of a nonparametric Mann–Kendall test. Table 6.22 shows the Mann–Kendall statistics (z-value) and related P value for precipitation, discharge, and maximum/minimum temperature time series.

From the Table 6.22, it is clear that there is negative trend at 5 and 1 % significance levels for precipitation at the two stations, Barcelonnette and Allos-Nord, and a positive trend for maximum temperature series at the Barcelonnette station. The trend analysis determined that the null hypothesis (H_0) is rejected in precipitation (at the Barcelonnette and Allos-Nord stations) and maximum temperature time series (at the Barcelonnette station), because the absolute value of Z is greater than $Z_{1-p/2}$, where $Z_{1-p/2}$ is obtained from the standard normal cumulative distribution tables.

This means that there is a significant downward trend in the precipitation time series and a significant upward trend in maximum temperature at Barcelonnette station. However, the P value of monthly discharge was greater than the significance level $\alpha = 0.05$ and one cannot reject the null hypothesis (H_0). This means that there is no significant trend in the discharge series (Table 6.16).

The obtained results, in this section, confirm the results of the IPCC in Europe which indicate a decreasing trend in precipitation and an increasing trend in

Table 6.22 Statistics of the Mann–Kendall test for selected variables

Parameter	Station	P value (5 %)	Kendall's tau	P value (1 %)	Kendall's tau
Precipitation	Barcelonnette	0.023	−0.063	0.023	−0.063
		H_a: There is a trend in the series			
Discharge	Barcelonnette	0.192	−0.036	0.192	−0.036
		H_0: There is no trend in the series			
Max-temperature	Barcelonnette	0.024	0.062	0.024	0.062
		H_a: There is a trend in the series			
Min-temperature	Barcelonnette	0.739	0.009	0.739	0.009
		H_0: There is no trend in the series			
Precipitation	Jausiers	0.379	−0.024	0.379	−0.024
		H_0: There is no trend in the series			
Precipitation	Uvernet	0.414	−0.021	0.414	−0.021
		H_0: There is no trend in the series			
Precipitation	Condamine	0.105	−0.044	0.105	−0.044
		H_0: There is no trend in the series			
Precipitation	Allos-Nord	0.054	−0.060	0.054	−0.060
		H_0: There is a trend in the series			
Precipitation	Sain-Paul	0.512	−0.021	0.154	−0.021
		H_0: There is no trend in the series			

H_a: There is a trend in the series (as the computed p-value is lower than the significance level $\alpha = 0.05$, one should reject the null hypothesis H_0, and accept the alternative hypothesis H_a). H_0: There is no trend in the series (as the computed p-value is greater than the significance level $\alpha = 0.05$, one cannot reject the null hypothesis H_0)

temperature. Thus, it could be concluded that the climate change signal has also been observed in the Barcelonnette area in a statistical point of view. Climate change does not depend on the local climate variability of the area, and it is better to consider more meteorological measurement stations to detect climate change in the wider region.

For graphical illustrations and detection of abrupt points in the precipitation time series at two stations, Barcelonnette and Allos-Nord, and for maximum temperatures at Barcelonnette station, the sequential Mann–Kendall test is presented in the Sect. 6.12.4.

6.12.4 Sequential Mann–Kendall Test

To see the changes in the trend over time, sequential values, $U(t)$ and $U'(t)$ were applied, from the progressive analysis of the Mann–Kendall test. $U(t)$ is a standardized variable and $U'(t)$ are computed similarly to $U(t)$ but only backward, starting from the end of the series. The linear regression model was used to estimate time trend of selected variables. The results show that all trend lines are significant at a 5 % level, according to the slope of linear regression for a precipitation decrease and a maximum temperature increase. In this section, the results of the sequential Mann–Kendall test for the period from 1961 to 2009 are presented. Figures 6.73, 6.74, and 6.75 show sequential values of $U(t)$ and $U'(t)$ statistics for precipitation (Barcelonnette and Allos-Nord stations) and maximum temperatures (Barcelonnette station). Horizontal dashed lines are corresponding to confidence limits at a 95 % significant level.

Inspecting the plot $U(t)$ of precipitation values from 1961 to 2009 (Barcelonnette), as shown in Fig. 6.73, two apparent decreasing significant trends are indicated. The first occurred between 1991 and 1993, and the second Sharp decline trend started at the end of the 2007 time series, which was significant at that time. $U(t)$ and $U'(t)$ intersect each other at several points and although they do not clearly intersect each other, some functions begin to diverge, indicating the starting point of an abrupt change. Based on the results, the sequential Mann–Kendall test showed evidence of individual abrupt changes in the magnitudes of precipitation from 1981 to 1982 and from 2001 to 2003.

Inspecting the plot $U(t)$ of the precipitation time series for the period of 1971–2009 (Allos-Nord) as shown in Fig. 6.74, two apparent decreasing significant trends are identified. The first one occurred between 1991 and 1993 and the second one started in 2009. $U(t)$ and $U'(t)$ intersects each other at several points and although they do not clearly intersect each other, some functions begin to diverge, indicating the starting point of an abrupt change. Based on the results, the sequential Mann–Kendall test showed evidence of individual abrupt changes in magnitudes of precipitation from 1977 to 1981 and in 2009.

Figure 6.75 shows the results of the forward/backward application of the sequential Mann–Kendall test for the maximum temperature time series at the

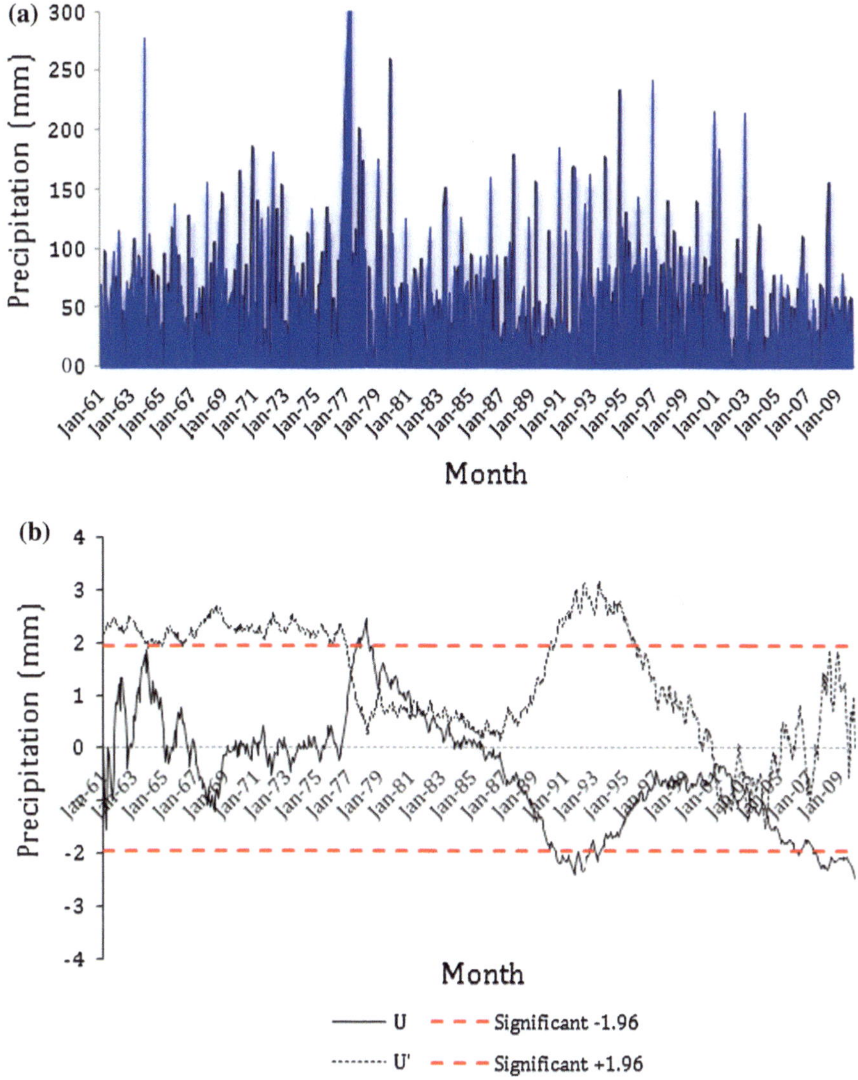

Fig. 6.73 **a** Normal time series of monthly precipitation; **b** sequential value of statistics $U(t)$ (*solid line*) and $U'(t)$ (*dashed line*) from the Mann–Kendall test for precipitation (mm) of Barcelonnette station, corresponding to the 95 % significant level (Barcelonnette station)

Barcelonnette station. The intersection point of forward and backward curves indicates the starting time of a sharp increase in the maximum temperature time series that started in 1994. After 2005, the trend of maximum temperature becomes significant. Abrupt or jump points are distinguished in 1993, 1994, and 1995.

From the results, it could be concluded that although there is a significant downward trend in precipitation (Barcelonnette station) and inversely, an upward

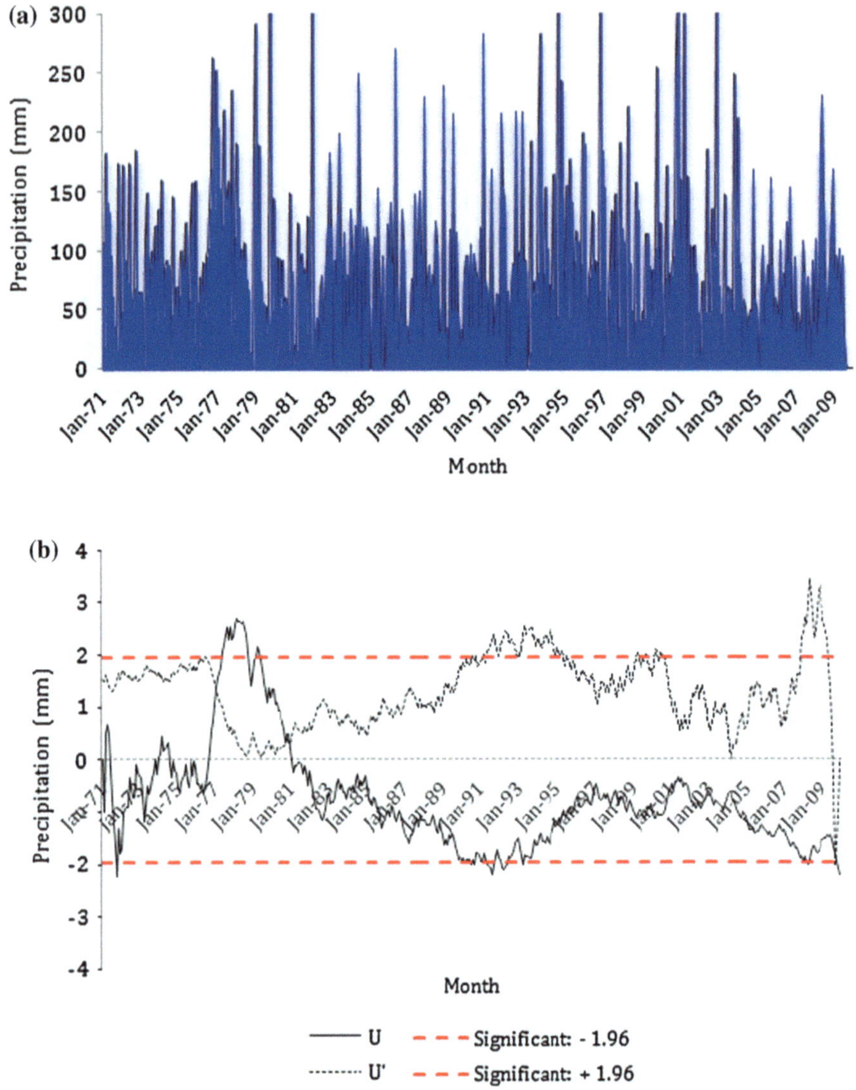

Fig. 6.74 **a** Normal time series of monthly precipitation; **b** sequential value of statistics *U*(*t*) (*solid line*) and *U'*(*t*) (*dashed line*) from the Mann–Kendall test for precipitation (mm) of Barcelonnette station, corresponding to the 95 % significant level (Allos-Nord station)

trend in maximum temperature (Barcelonnette station), there is no significant trend in observed discharge data (as mentioned in Sect. 6.12.3).

A downward trend in precipitation and an upward trend in temperature push the researcher to find other triggering factors which affect normal distribution of discharge in the Ubaye River. Due to the special position of the region, the Barcelonnette area enjoys a Mediterranean climate affected by a mountain climate

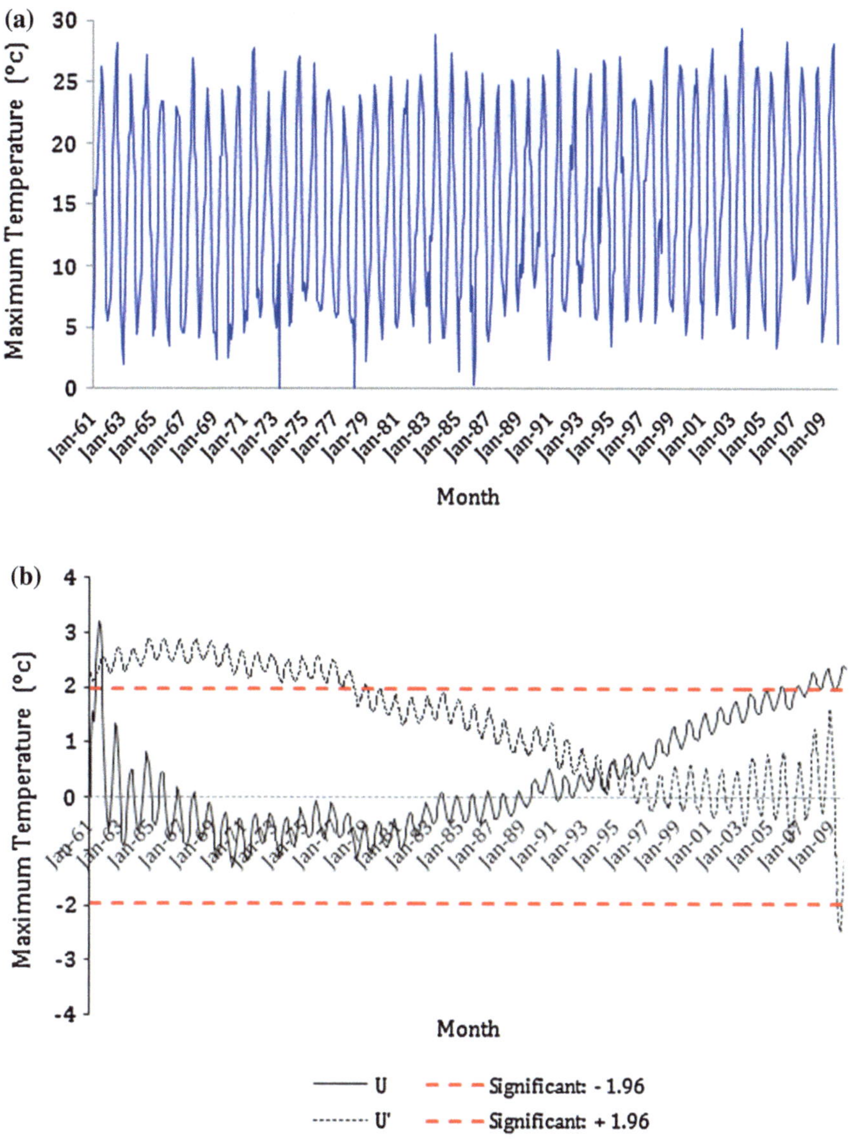

Fig. 6.75 **a** Normal time series of monthly temperature; **b** sequential value of statistics $U(t)$ (*solid line*) and $U'(t)$ (*dashed line*) from the Mann–Kendall test for maximum temperature (°C) of Barcelonnette station, corresponding to the 95 % significant level

and has a mild continental influence. This effect of the Mediterranean climate can be seen by the impact of warm and dry winds called Sirocco from the south, which blow over the entire region resulting in high snowmelt in the spring and high discharge in the Ubaye River. Therefore, as Weber mentioned in his investigation [2], it should be considered that the Barcelonnette Basin is affected by the

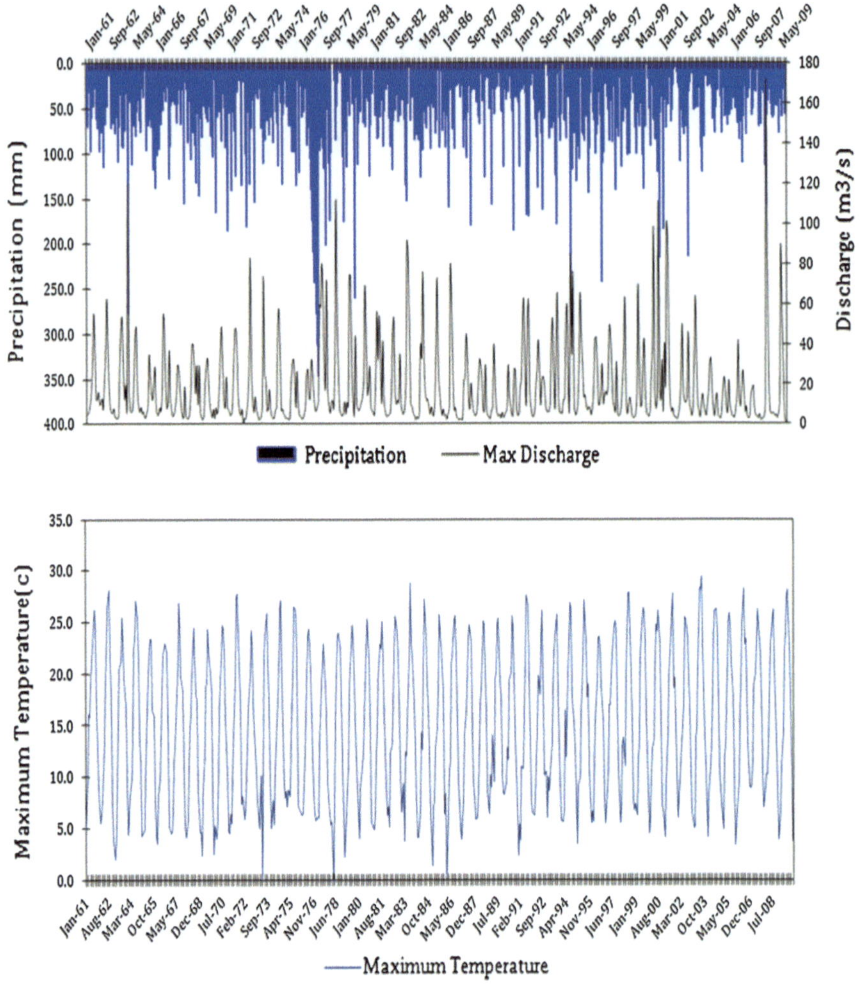

Fig. 6.76 Monthly precipitation, discharge, and temperature in Barcelonnette station for the period of 1961–2009

Mediterranean marine influence moderated by a mountain climate and a mild continental influence. The orographic influence causes precipitation to increase with elevation, typical in the form of an "attenuated Mediterranean regime".

In time series analyses, two maximum peaks for precipitation in September 1963 and August 1978 can be realized. It is noteworthy to point out that there was a decrease at this maximum temperature as well as a decrease in the amount of discharge at the same time. Humidity in this period may have been high, which is affected by winds and rainfall. Humidity depends on water vaporization and condensation, which, in turn, mainly depends on temperature.

Figure 6.76 shows monthly precipitation, discharge, and temperature at the Barcelonnette station for the period of 1961–2009. As was observed in the

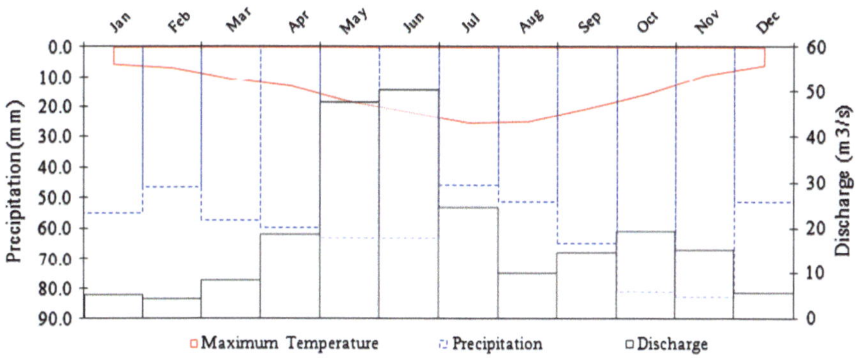

Fig. 6.77 Average monthly precipitation, discharge and temperature between 1961 and 2009

previous results, monthly precipitation has experienced a minimal decline. Monthly maximum discharge, on the other hand, is relatively stable. Furthermore, runoff does not increase with a simultaneous increase in precipitation; and the rate of changes in precipitation is not necessary equal to the amount of runoff. For instance, while precipitation dropped from 2007, runoff increased.

As partly conclusion, as mentioned in pervious sections, on the monthly scale, the results of *p* values statistical analysis show that only precipitation and maximum temperature are statistically significant with *p* values of 0.023 and 0.024 at significant levels of 1 and 5 %. On the other hand, the *p* value for discharge represents non-significant levels of both 1 and 5 %. The Mann–Kendall test analysis shows there is a negative and downward trend in precipitation and for maximum temperatures and a positive or upward trend with Kendall's tau of −0.063 and +0.062, respectively.

So far, all achieved results force the research to focus more on monthly and daily available data. Figure 6.77 represents the average monthly precipitation, discharge, and temperature between 1961 and 2009.

Daily analysis (maximum snow depth, maximum discharge, precipitation, and maximum temperature) between 1995 and 2005 is also shown in Fig. 6.78. The results of analysis show that during the summer the rate of snowmelt has a direct impact on the rivers peak discharge. In both flood events of 1957 and 2008, the rate of snowmelt was higher than the normal rate in other years.

Bhattacharya [1] interviewed local people and, according to them, there has been neither a decline in the amount of snowfall nor have there been more frequent wetter winters and drier summers, which is a possible indication of climate change in the area.

The results of cross-correlation (CCF) (snow depth–discharge) also show that there is a strong backward negative correlation between maximum discharge and snow depth. When snow packs start to melt this will affect the rate of runoff in the Ubaye River (Fig. 6.79).

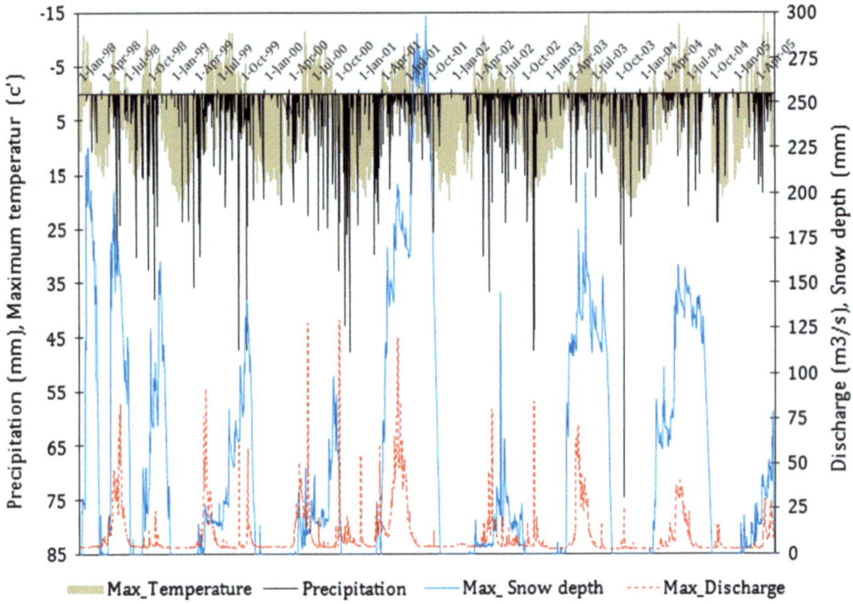

Fig. 6.78 Daily maximum snow depth, maximum discharge, precipitation, maximum temperature

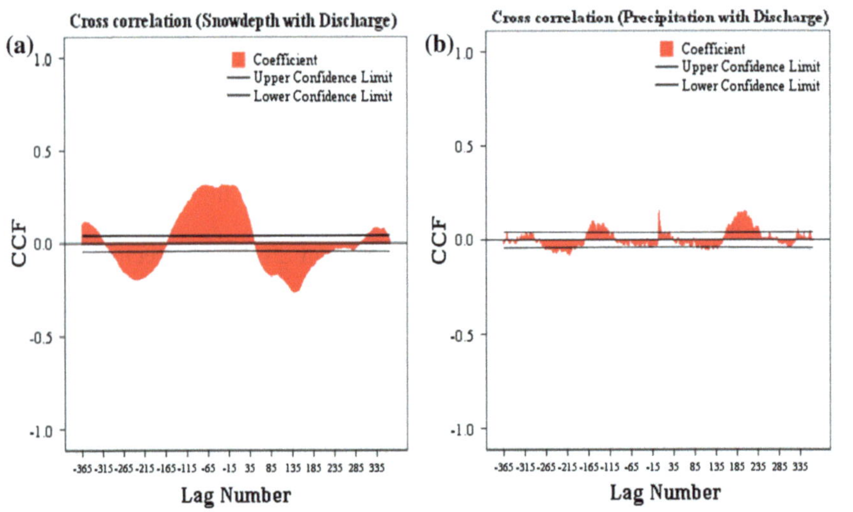

Fig. 6.79 Cross-correlation (CCF). **a** Snow depth and discharge. **b** Snow depth and precipitation

6.13 Land Use Change Modeling

In Chap. 2, land use change up until the twentieth century was discussed according to the available literature. The photographic archives of the mountain terrain restoration (MTR) agency provided views of the watershed dating as far back as 1860. This part deals with detecting land use change during 1956–2004 based on the available aerial photos.

The current land use categories in the study area are mostly classified into coniferous forests (~55.94 %), broad-leaved forests (~4.85 %), natural grasslands (~0.8 %), arable land/permanent crops (~13.64 %), pastures (~8.38 %), bare rocks (~1.6 %), black marls (~2.86 %), urban fabric (~11.24 %), airdrome (~0.45 %), water course (~0.21 %), marshes and water bodies (~0.01 %).

In this study, the evolution of watershed land use was successively studied and compared using aerial photo interpretation methods and GIS applications. The results show that the total forest area of the case study region decreased from 46.26 to 51.65 % during the period of 1956–2004, which shows a transformation from forest to urban areas. Arable land areas decreased from 12.86 to 5.86 %. Conversely, urban areas increased from 1.67 to 4.01 %. Later, some parts of these bare lands which were closer to the city were transformed into urban areas where mass housing was clearly seen. Figure 6.80 classifies different land use maps according to aerial photos and Table 6.23 shows statistical results of classification classes 1956, 1974, 1982, 1665, 2000, and 2004.

The period of 1956–2004 shows a huge transformation of arable lands into other land use types. Arable land has continued to enlarge and be transformed into urban land. The urban growth rate was ~4.01 % and deforestation characterizes

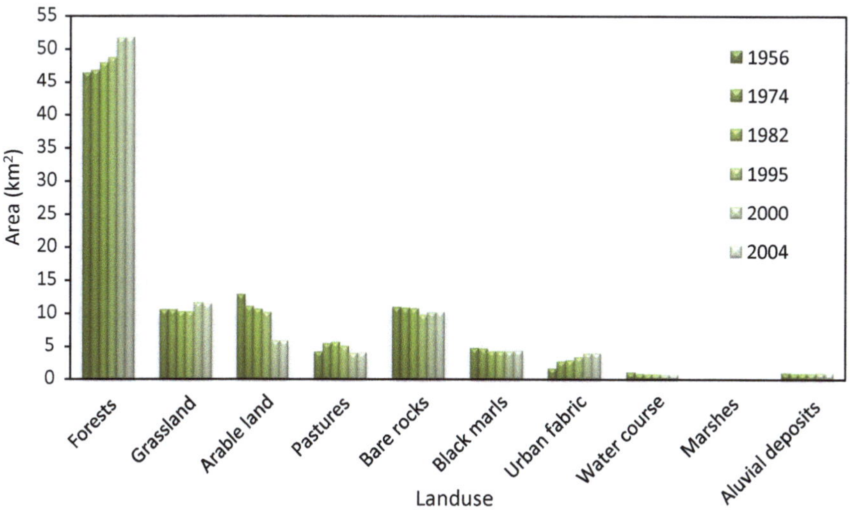

Fig. 6.80 Land use changes during 1956–2004

Table 6.23 Calculated area (km^2) for different land use change over time

Type	1956	1974	1982	1995	2000	2004
Coniferous forests	46.26	46.68	47.83	48.60	51.54	61.65
Natural grassland	10.51	10.43	10.22	10.22	11.51	11.35
Arable land	12.86	11.01	10.54	10.08	5.79	5.86
Pastures	4.20	5.40	5.63	5.03	3.93	4.04
Bare rocks	10.88	10.79	10.67	9.71	10.14	10.07
Black marls	4.76	4.66	4.29	4.24	4.24	4.37
Urban fabric	1.67	2.70	2.87	3.40	3.98	4.01
Water course	1.03	0.82	0.77	0.75	0.66	0.65
Marshes and water bodies	0.02	0.03	0.03	0.03	0.04	0.04
Alluvial deposits	0.99	0.82	0.86	0.89	0.90	0.84

this period. New construction has played an important role in encouraging urban expansion and population increase. GIS-based analysis results show land use changes from 1956 to 2004 (Fig. 6.81). These findings confirm anthropogenic effects over the case study area and could be considered as main deriving factors in the river's regime changes and also as an accompanying factor of climate change in this study. This point was discussed in the climate change sections in more detail.

Figure 6.82 shows more focusing on Barcelonnette area and compare the land use change in 1956 and 2004.

Figure 6.83 shows the land use change over time: (a) 1956, (b) 1974, (c) 1982, (d) 1995, (e) 2000, and (f) 2004. The red-dash line represents decreasing arable land and increasing urban areas over time.

More focusing on results, as shown in Fig. 6.84, confirm this fact that the arable lands, the forests, the urban areas, and the water course were faced the most changes from the past to now.

The result of this investigation is shown in Table 6.24.

Being a touristic center, Barcelonnette has faced rapid urbanization over the past 43 years. As part of these developments, this study provides important insight into the urban land use changes that took place between 1956 and 2004, and aids in monitoring these changes and in assessing the effects of planning decisions. An integrated approach of GIS has been shown as an effective technique for the study of land use analyses. The spatial analysis of thematic information which can be derived from GIS helps in the assessment of developmental plans before they are implemented in the field. For a better comparison of the changes between 1956 and 2004, the results are shown in Figs. 6.85 and 6.86.

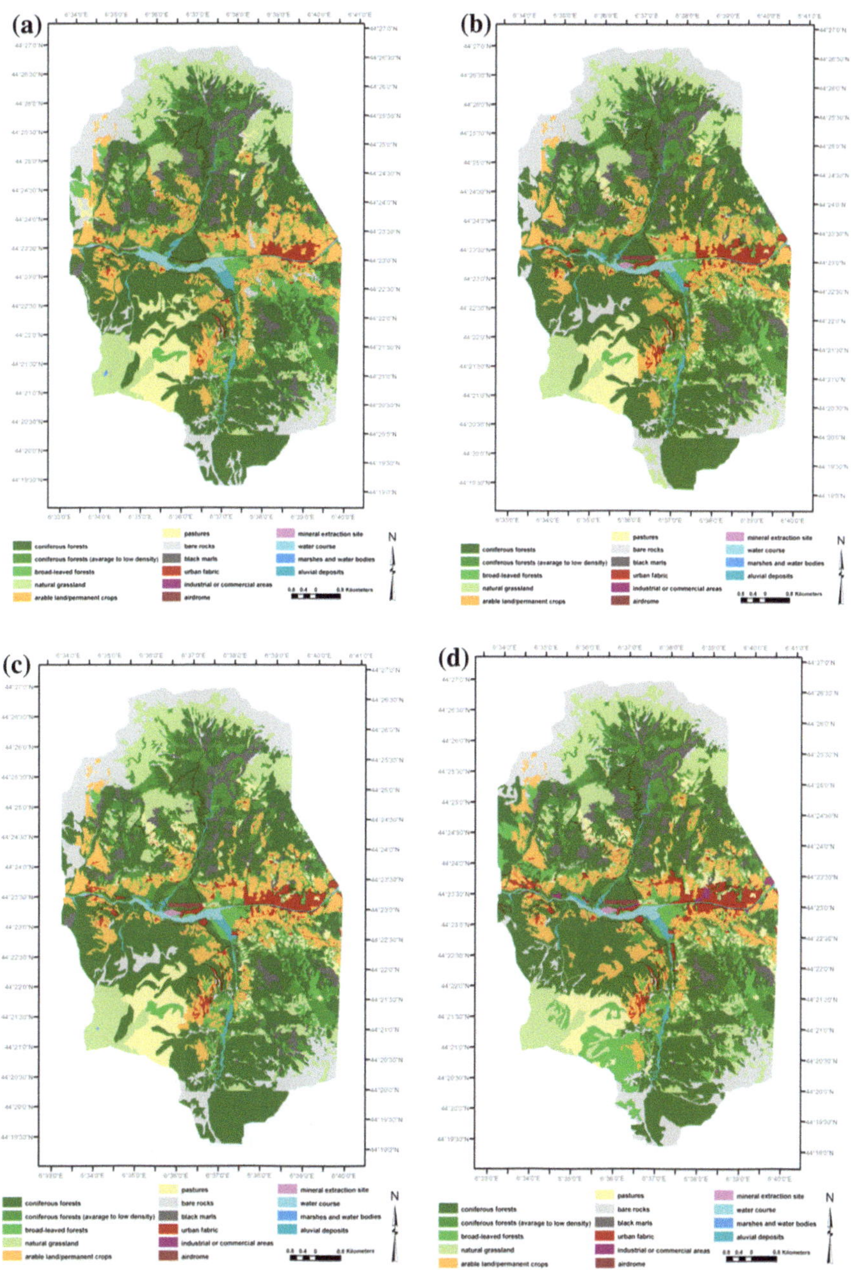

Fig. 6.81 Land use change over the time; **a** 1956; **b** 1976; **c** 1982; **d** 1995; **e** 2000; **f** 2004

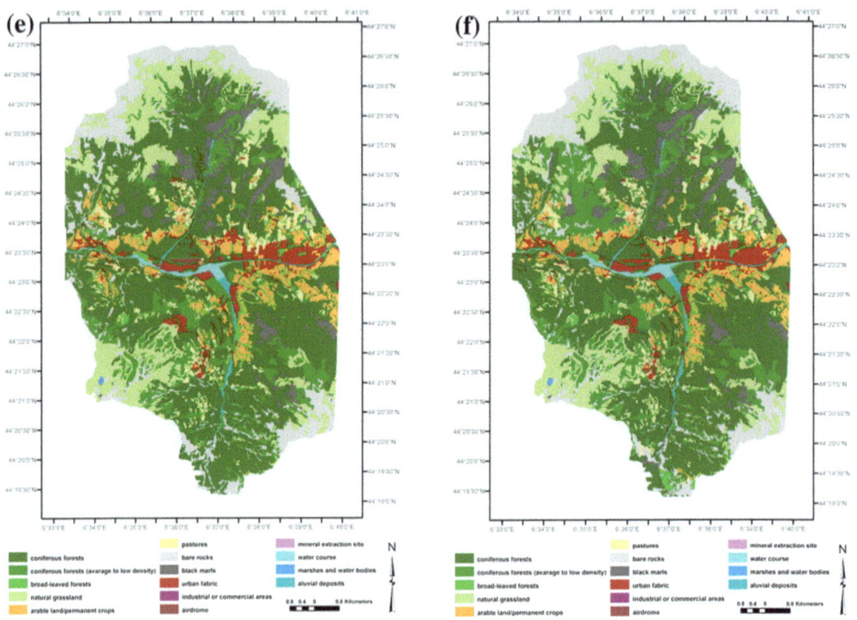

Fig. 6.81 continued

Fig. 6.82 Land use change in 1956 and 2004 (Barcelonnette area)

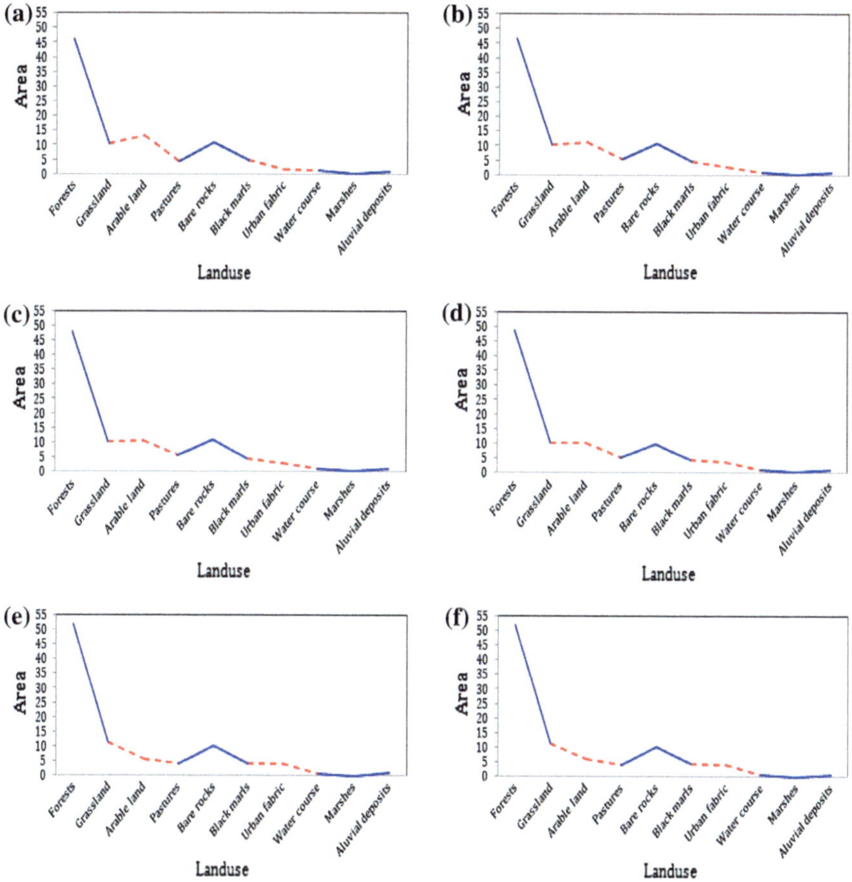

Fig. 6.83 Land use changes over time (area km^2). **a** 1956, **b** 1974, **c** 1982, **d** 1995, **e** 2000, **f** 2004; the *red-dash line* represents decreasing arable land and increasing urban areas over the time

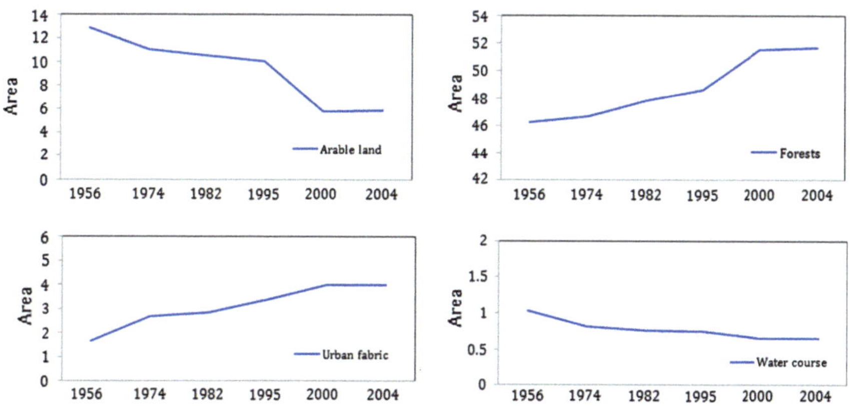

Fig. 6.84 Changes in area (km^2); arable land, forest, urban area, and water course during 1956–2004

Table 6.24 Different land use change over time in percent (%)

Type	1956	2004	Percent (%)	Trend
Forests	46.2	51.6	12	Increase
Arable land	12.8	5.86	−54	Decrease
Urban area	1.67	4.01	140	Increase
Water course	1.03	0.65	−37	Decrease

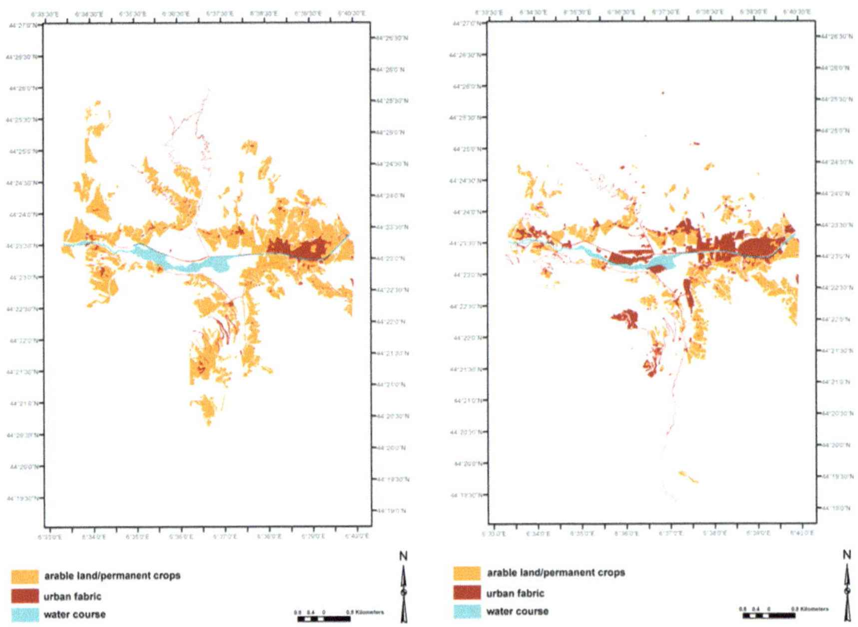

Fig. 6.85 Comparing arable land and urban areas in 1956 and 2004

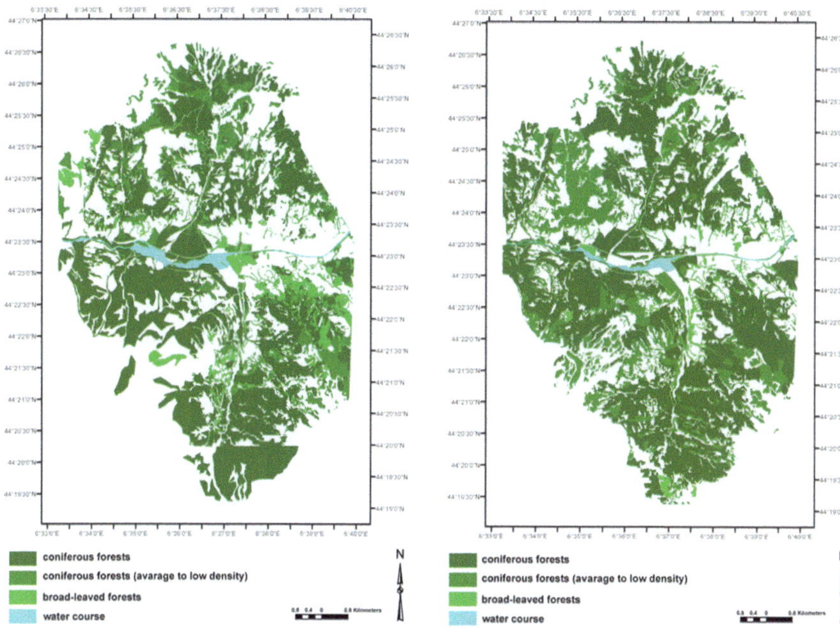

Fig. 6.86 Comparing coniferous forest in 1956 and 2004

6.14 Chapter Summary

In hydraulic modeling, the output is affected to a large extent by model input parameters. For input data preparation, the quality of the DEM was at first controlled using interpolation methods as important input data to accomplish successful modeling. The results showed that the quality of the original DEM was sufficient enough for modeling. The hydrological tool in ArcGIS was applied only to fill the probable sinks in DEM. In the next step, the original DEM was considered as the reference data source, and then different interpolation techniques were compared to access the best method to interpolate DEM based on observed data points. For this purpose, different statistical analysis methods including Q–Q plot and spatial analysis were applied to estimate the errors.

Before starting hydrodynamic modeling, based on the available data, land use maps from 1956 to 2004 were analyzed to gain better insight on river morphology changes over time and to access a variety of roughness derived from different land use maps to assign the cross sections in the flood modeling phase. Then, the focus was on preparing hazard maps and frequency analysis. The main purpose here was to compare some severe flood events that occurred in this area and to estimate flood propagation using hydrodynamic models (One dimensional (1D) HEC-RAS model and one-two dimensional (1D–2D) SOBEK model), with respect to a variety of cross sections, river morphology and different hydrographs. Moreover, the preparation of different hydrological scenarios for civil protection purposes was considered in the study area, including: (1) DEM scenario—the model was run with different DEM resolutions to understand how the grid resolution of a DEM affects the flood characteristics when simulated by a 1D and 1D–2D flood models; (2) river changes scenario—the aim was to evaluate river change scenarios in order to know how river channel changes affect the flood, using different profiles and different discharge values; (3) dyke failure scenario—the major aspect was realizing the possible effects of dyke failure before a severe flood phenomenon and finally (4) model scenario—was used to compare two different hydrological models including HEC-RAS (1D) and SOBEK (1D–2D), in order to know what are the advantages (or strengths) and disadvantages (or weaknesses) of these models. In the last part of the hydrodynamic simulation, the probability distribution and the return period were analyzed.

Climate change is one of the most important triggering factors for a flood event. The historical trend analysis of hydrological and meteorological series is important and even more relevant when considering the regional effects of global climate change. Therefore, to detect climate change and analyze the climatic parameters, any probable abrupt points and trend estimations in the time series of climatological and hydrological parameters using statistical methods were explored. The overall conclusion from this section revealed that there is an increasing trend in temperature time series and an inversely decreasing trend in precipitation time series obtained from recorded data in the stations located in the Barcelonnette basin. The distribution of discharge in the channel was normal and no statistically significant trend was distinguished in flow discharge at the Barcelonnette station.

Additionally, to detect the land use changes from 1956 to 2004, first the land use maps were digitized using GIS techniques based on the available aerial photos. Then, the results were analyzed statistically. From the results, it was realized that the urban areas increased about $\sim +140\ \%$ and arable land showed a decreasing trend ($\sim -54\ \%$) from the past until the present. Results over forest areas also represented an increasing trend ($\sim 12\ \%$), inversely due to channelization and the decreasing human activity in the channel ($\sim -37\ \%$) over time.

To successfully provide scientific feedback on the relationship between climate change, land use change, and flood events in the Barcelonnette area, all achieved results were reviewed and the overall conclusion and final discussion is presented in the Chap. 7.

References

1. Bhattacharya N (2010) Flood risk assessment in Barcelonnette, France, M.Sc. Thesis, International Institute for Geo-information Science and Earth Observation (ITC), Twente University, Enschede, The Netherlands
2. Weber D (1994) Research into earth movements in the Barcelonnette basin. In: Casale R, Fantechi R, Flageollet JC (eds) Temporal occurrence and forecasting of landslides in the European Community, Final report, vol I. Contract EPOCH, European Commission, pp 321–336

Chapter 7
Overall Conclusion and Discussion

7.1 Overall Conclusion

The methodology applied in this research aimed to find a scientific relationship between flood events, climate change, and land use change from the past until the present. The objectives of the dissertation were to develop and apply methods for flood hazard assessment, to detect land use change as well as climatological and hydrological trend analysis in time series. As an overall conclusion, this chapter attempts to review and discuss the research objectives based on all achieved results:

Objective (a):

- As for the first assumption, interpolation methods that often assume data points in original DEM are correct (exact), but it can be assumed that they are subject to error (generally of a known or estimated extent). This means the interpolation technique is not only for creating DEM from contour lines or observed data points, but also for correcting DEMs that contain errors, such as sink errors. Therefore, the main purpose was to correct the original DEM, find the possible errors, and to address the best interpolation techniques in order to erase these errors. To compare the accuracy of the original DEM and modeled DEMs, the stream network was derived for each interpolated surface and the original DEM. By visualization analysis of all achieved results, it seems that IDW, Spline, Circular, Exponential, and Stable produced the poorest results mainly in the upstream, where a small part of the main channel is missing. The results obtained from Topo to Raster, Gaussian, and Spherical-derived drainage networks coincided better with the reference stream network. Still, there is some discrepancy in the modeled network in some parts where the river follows a

A. Ramesh, *Response of Flood Events to Land Use and Climate Change*,
Springer Theses, DOI: 10.1007/978-94-007-5527-7_7,

wrong path. As an overall result, it was concluded that the interpolation techniques were not successful in creating a perfect DEM more accurate than the original DEM. Consequently, the first assumption is rejected. Therefore, in the hydrodynamic simulation the original DEM was applied. The quality of DEM was also tested hydrologically using the drainage enforcement algorithm to recognize that each spurious sink is surrounded by a drainage divide containing at least one saddle point.

- The second assumption was that the original DEM is included in observed elevation data points with 100 % accuracy. Therefore, it was used as a correct source to test the different interpolation methods. The idea behind this assumption was to find the best interpolation technique which represents the best agreement with the observed elevation of the data point set. As a general conclusion, the results of analysis showed that all DEMs derived from different interpolation techniques were statistically significant. Going into the details in terms of RMSE, R-square, maximum absolute error and frequency analysis, better results in model accuracy were obtained when Exponential, Gaussian, Stable, and Spherical models were used instead of Circular, IDW, Topo to Raster, and Spline. Therefore, as an overall result, it could be concluded that Geostatistic techniques show the strongest results compared to deterministic. In between, the Spherical showed better agreement with the observed data and represents the smoothest and the most accurate DEM. The Q–Q plot was also applied to quantify and represent the estimated variable's distribution against the original values more consistent for Geostatistic techniques. It could be concluded that in Geostatistic methods the distribution of a variable matches very strongly with the observed or original data distribution. On the other hand, the points cluster around a straight line. "Residuals" analysis showed that the best results also belong to Geostatistic-Spherical methods, which achieved the most accurate results with a value of $<.001$. Consequently, it could be concluded that Geostatistic techniques show the strongest results compared to the deterministic. In between, the Spherical and Gaussian showed better agreement with the observed data and represents the smoothest and more accurate DEM. In terms of spatial analysis, the standard residual diagnosis graphs shows that the best results in terms of normal distribution belongs to Gaussian method because all scattered values spread close to zero (skewness is 0.071) between the ranges of −0.5–0.5. As an overall conclusion the residuals data set in Gaussian method obeys a normal distribution. The results of Moran's I investigation showed Gaussian and Stable showing perfect dispersion in the first distance. On the other hand, the Moran's I value for IDW, Topo to Raster, Spline, Circular, Exponential, and Spherical represent perfect correlation in the first distance. Inbetween the lowest Moran's I value of 0.013 belongs to Spherical. To compare the accuracy of the interpolated DEMs, the stream network for each surface was derived using ArcGIS. The modeled stream network was compared with the stream network digitized from topographic maps ("true" data). By reviewing the figure it seems that the overlay of (b) IDW, (c) Spline, (d) Kriging, Circular, (g) Kriging, Exponential and (h) Kriging, Stable produced the weakest results, mainly in the upstream where a small part of the channel was missing. The results obtained from (e) Topo to Raster, (f) Kriging, Gaussian and (i) Kriging, Spherical are considered to be in better agreement with the observed drainage network. By focusing on details and the interpretation of results, it is concluded that the drainage networks generated by the Kriging, Spherical allowed a much more accurate delineation of the stream network. Focusing more on details, as would be expected, the Spherical was the best overall modeled drainage network, and it was comparatively more accurate than the river from the DEMs generated with other interpolators, even in the floodplain which is a flat area.

Objective (b):

Within hydrodynamic modeling, the output is affected to a large extent by model input parameters such as cross-section, Manning's value, etc. Therefore, before starting the hydrodynamic simulation, it was necessary to prepare all input data as follows:

- First, the Ubaye River was divided into three parts: (1) the upstream or channelized part of the river in the city area; (2) the Middle part of the river in the natural environment when the river morphology starts to widen; and (3) the downstream part of the river in the natural environment when the river starts to narrow again. Cross-sections should be added at locations where changes occur in slope, width, and roughness. Therefore, to select the cross-section's location with respect to all the above mentioned parameters, 10 cross-sections were applied along the river and the data was manually imported to the SOBEK (1D–2D) model. The interval between cross-sections was selected at 150 m, with a total river section of 10 km in length in the HEC-RAS (1D) model. All the cross-sections were created based on a 5 m resolution DEM using the HEC-GeoRAS model. In regard to the modeling representing any prefect results, it should be noted that these cross-sections also needed to be manually modified before being used in flood simulations (SOBEK and HEC-RAS).
- While using Manning's equation, the selection of a suitable value of n is the most important single parameter for the proper estimation of velocity in an open channel. To be sure about the correct selected roughness value for the channel, the model was run for different values from 0.02 to 0.07. The result of hydraulic depth was compared with observed data at the Abattoir Bridge. The testing results show that the friction value of 0.05 in the channel represents the best agreement between modeled and observed water height measured close to the Abattoir station.
- Due to a lack of available recorded data over the case study area, calibration analysis was limited to the observed height of water levels found in the literature and field reports of the 2008 flood. The adjustment was performed using different friction values in the channel. Considering the lack of data regarding the river morphology, the model was considered to have been a successful calibration.

In order to have valuable and basic information available for any future management, it is necessary to provide the authorities, stockholders, other scientists, and the public with basic flood hazard maps which contain fundamental information for any environmental, technical, economical, and political decisions regarding flood event management. This shows the potential adverse consequences associated with different flood scenarios for civil protection purposes. The results of different scenarios are discussed as follows:

DEM scenario:

- The DEM must be a continuous surface that includes the bottom of the river channel and the floodplain to be modeled. Because all cross-sectional data should be extracted from the DEM, only high-resolution DEMs accurately representing the ground surface should be considered for hydraulic modeling. The calculated water velocity by SOBEK and HEC-RAS models shows that this hydraulic parameter decreases with increasing DEM pixel size. This is related to the fact that the model considers more roughness and friction as it uses the higher resolution DEM. A major disadvantage of using the low resolution DEM is the loss of some features such as dykes and levees. These features play an important role on flood model simulation results. Visualization analysis shows that, due to a low resolution of 10 m, compared to 5 m, the generated river network for 10 m deviates more and does not agree well with the actual drainage paths. But with the increase of horizontal and vertical resolutions, the delineated river networks from 5 m DEM come in close agreement to the actual flow path. Even in flat urban areas, the results of 5 m resolutions are more satisfactory compared to 10 m resolutions. The results of the 10 m DEM simulation show that the flow distribution shifted from the actual flow paths in locations of the generated river network. It was obvious that the flood mainly occurred due to overflow in the rivers, which was perhaps due to the negligence of the locally elevated lands such as dykes and roads in the model.
- The idea behind resampling the DEM was to estimate how much the resampled data were close to the data source. As a general conclusion, in terms of RMSE analysis, better results in the accuracy of the model were obtained when a DEM was used after resampling. The mean error in resampled DEM was 4.292, much less than the value of 25.928 for DEM before resampling. Therefore, as an overall result, it could be concluded that a low resolution DEM could show better and more accurate results when it is resampled to a higher resolution.

River Change Scenario

- Regeneration of past events was the major concern in this research. But with the well-calibrated model based on observed data for the 2008 flood event, the regeneration of other past events was possible and convincing enough. The results were validated by available photos from the past event. The error behind the results is due to a lack of high spatial and temporal data resolution, especially in hourly discharge and in the correlation between observed hourly water height and discharge.

- The digital elevation model that was used in the 2008 flood event was adjusted by increasing the heights of the dyke by 0.5 m from that of the 1957 event. The maximum estimated velocity for the 2008 flood in the channel was about ~3 m/s and had a maximum water depth of 2.7 m. At cross-section 12, located exactly at the Abattoir Bridge, the maximum water depth in the near bank was 1.7 m. This result shows the best agreement between measured and observed data. This means that the model calibration was done in the right way to access more accurate and real results.
- Flow simulation results for 1982, 1995 and 2000 offered an idea about the area and river change and show critical places in flow propagation of velocity and water depth. River changes between 1957 and 2008 show that the river became narrow especially in the middle. The dyke height in the city area also increased by 0.5 m after the 1957 flood, preventing the urban area from any probable flood damages. In all simulations, the right bank was much more sensitive to the velocity of water. No inundation was observed in natural part of the river. Probably, the vegetation on the banks has a small contribution to overall flow resistance.

Dyke scenario

- In case of dyke failure, it is essential to evaluate the possible effects in advance. In this scenario (1), normal retention water depth and velocity were shown in the channel—water flow was kept between the banks. The model indicated that the maximum water depth and velocity in the Ubaye River were approximately ~2.7 m and ~3 m/s, respectively. This corresponds to the field reports of the 2008 flood. The idea behind scenario (2) was to test the capacity of the channel to the same hydrograph of the 2008 flood in case of dyke failure. The results showed that the flow propagation was affected by the channel maintenance and structural dyke integrity, and in several locations the flow overtopped the bank and the floodplain was affected by inundation. The assumption was that the lack of maintenance produced the dyke break and inundation even with a low amount of water. Scenario (3) was used only as support scenario to compare the simulation with other scenarios. The results obtained from scenario (4) showed that the dykes overtopped on several sections in both banks, especially in the urban sectional area. Critical overtopping occurred between cross-sections 11 and 12 on both the left and right banks (~2.9 km from upstream). This shows that at the current river morphology and dyke structure, although this hydraulic structure was a good strategy to protect the city part of the river in case of severe flood, there is still concern about overtopping and inundation in case of dyke failure.

Model Scenario

- In this study two different flood modeling software packages were used to simulate floods. Based on the user's purpose, it has to be determined which model should be applied. If engineers have to decide on dams or house enforcements against flash flows, the SOBEK results are more suitable due to higher accuracy of the flow velocity results. However, if only the approximate extent of water is important (e.g., for spatial planning procedures), the HEC-

RAS model is sufficient. In general, combining the results of modeling both in 1D and 2D domains is necessary. The results of flood propagation in both models gave us better insight into flood characteristics such as, where the water moves (inundation) and how fast the water flows (water velocity). These results are very important as basic information for policy making, decision support, and flood hazard planning.

Return period

- The results of statistical methods for frequency analysis of the Ubaye River in Barcelonnette show that the 1957 flood event may, on average, be expected annually every 11 years with a 90 % lower confidence level, and every 66 and 100 years with a 90 % upper confidence level. In addition, a simulated flood hazard for a 66 year return period has a 0.015 probability of occurrence; and 100 year return period has a 0.01 probability of occurrence.

Objective (c):

- In case of a maximum discharge of 480 m^3/s the velocity and water depth increases in the channel, between the left and right banks in the channelized part in the city and in the undisturbed river bed. But critical overtopping of the dyke is observed on both left and right banks in the city area. It shows that with current morphology of the river and current dyke structure it could protect the area in case of such designed flood events.
- In a natural environment, the vegetation covering the banks might typically be a small contribution to the overall flow resistance in the natural reach of the river. Restoring vegetation along the river would increase hydraulic resistance during flood events. Higher bed roughness equates to lower near-bank velocity. Restoring vegetation along the river would increase hydraulic resistance.

Objective (d):

- As an overall conclusion on climate change detection, a significant downward trend in the precipitation time series at both the Barcelonnette and Allos stations were distinguished, while an inverse upward trend in the maximum temperature time series at Barcelonnette was identified. The distribution of discharge in the channel was normal and no trend was realized in the flow discharge at the Barcelonnette station. A negative trend in precipitation and a positive trend in temperature push the research to find other triggering factors which affected the normal distribution of discharge in the channel.
- The results of the time series analysis between 1961 and 2009 showed that the highest precipitation occurred in 1977. Therefore, this year was selected to compare to 2008, the year which the recent flood was occurred. Monthly data of

precipitation discharge and temperature was analyzed for both years. The amount of precipitation for almost all months in 1977 was more than in 2008. In general, precipitation shows a decreasing at the end of the period (1961–2009). By investigating all achieved results from the time series analysis, it could be concluded that: (a) the amount of precipitation for almost all months in 1977 was more than in 2008. In general, precipitation shows a decreasing trend at the end of the period (1961–2009); (b) the temperature for almost all months in 2008 shows an increase from 0 to 4 °C and; (c) maximum discharge in 2008 was more than in 1977.
- Flood 2008 occurred with a decreasing amount of rainfall comparing to 1977, therefore, as an overall conclusion, it is assumed that the land use change and melting of snow are acceptable reason for the 2008 flood.
- The result of daily analysis for the time series (including maximum snow depth, maximum discharge, precipitation and maximum temperature) showed that during the summer, the rate of snowmelt has a direct impact on the rivers peak discharge. In both flood events of 1957 and 2008, the rate of snowmelt was higher than the average in other years.
- The non-homogenous precipitation time series at other investigated stations may be affected by multiple changes in the location of the gauge and may be one change of instrumentation.

Objective (e):

- From the results, it was realized that urban areas increased ($\sim$+140 %) and arable land showed a decreasing trend ($\sim$−54 %) from 1956 to 2004. Forests also represent an increasing trend ($\sim$12 %), inversely due to channelization and human activity, which caused the channel to decrease ($\sim$−37 %) over time.
- No inundation occurred when the 1957 flood was simulated for current river morphology, due to reconstruction and increasing height of the dyke as a constructive human activity in Barcelonnette City.

7.2 Discussion

According to all the achieved results of this study in the Barcelonnette area, this section attempts to review and discuss the hypotheses as pre-research predictions as follows:

Hypotheses (a): Interpolation techniques are suitable tools to control the quality of digital elevation model.

- In this research, applying the interpolation techniques was not necessary because the available 5 m resolution DEM was more accurate, precise, and had a higher density than the modeled DEM derived from different interpolation techniques.
- For future research, a LiDAR DEM would be more accurate, but SOBEK as 1D–2D modeling need weeks or months to calculate a small flood. In this case, the DEM needs to be degraded to a larger grid, or it is recommended to apply another hydrodynamic model to simulate a flood.

Hypotheses (b): For any flood hazard assessment, simulation of variability of flood crisis scenarios and flood frequency analysis is necessary for civil protection purposes.

- The results of the hydrodynamic section and a subset of different crisis scenarios are very important as basic information for policy making, decision support, and flood hazard planning in the Barcelonnette area. It seems that for any future flood planning, the flood hazard/risk zone provided by Prévention des Risques naturels (PPR), Prévention des Risques naturels (PER) and Restauration des Terrains de Montage (RTM), should be updated based on the results of this study. Additionally, the natural areas in this region also need to be considered for providing hazard and risk maps.
- Return period and flood frequency curves should be checked and updated from time to time.

Hypotheses (c): Topographic elements can be used as a strategic flood defense system, especially to delay flood waters in order to evacuate people.

- There is still concern about overtopping and inundation in the case of a severe flood event with the current dyke height. Therefore, it is recommended that the flat areas in both the city and natural sectors be protected with dykes. Maintenance and construction of the dykes should be controlled. It should be considered that the dyke can induce catastrophic floods when they start to fail. If dykes fail, they can cause a false sense of security on public and residential parts in flood plain areas of Ubaye River.
- In principle, all artificial structures need more and more maintenance with time; thus, the probability of failure continuously increases, which is why channel modifications/improvements/hydraulic structure improvements are foreseen.

Hypotheses (d): Climatic variables influence flow distribution.

- A decreasing trend in precipitation and an increasing trend in temperature and seasonal changes seem to be a sign of climate change. Especially, increasing temperature leads to more snowmelt during the spring. The results of cross-correlation for water depth and discharge also showed that there is a strong backward negative correlation between maximum discharge and snow depth. This means that as snow packs start to melt, the result of this melted snow will affect the rate of runoff in the Ubaye River.

Hypotheses (e): Flow propagation in the Ubaye river is affected by anthropogenic activities and land use change.

- The quick rate of expanding urban areas and the decreasing amount of arable lands affect the amount of discharge. This leads to more rainfall directly flowing into the channel, instead of infiltrating into the soil.
- Land use changes such as increasing deforestation and urbanization, elimination of floodplains area in Barcelonnette lead to reducing of infiltration capacity and increasing of runoff coefficient.
- Urban area was expanded on the floodplain of Ubaye River. Consequently the ability of the floodplain to attenuate a flood peak by promoting storage, infiltration, and alternative flow pathways was reduced.
- According to the results of flood hazard maps in this research, it seems that by floodplain development, increasing accumulation of population and in flood-prone areas, humans have been driven to occupy unsafe areas, thereby increasing the loss potential. Therefore, as long-term planning for the future, removing buildings from dangerous areas is recommended.

Finally, to finish the discussion, and as a consequence of this work, it could be assumed that climate change, land use changes, anthropogenic activities, and snowmelt are the most important factors in the fluctuation and changes in the discharge and flow regime of the Ubaye River. Therefore, there is a need for continuous monitoring of the channel and related hydraulic structures. In the next chapter the perspectives for future research are discussed.

Chapter 8
Perspectives

The outcomes of this research highlight the following directions for further investigation in the Barcelonnette area:

8.1 Risk Analysis and Management

In order to have an available and effective tool for information, as well as a valuable basis for priority setting and further technical, financial, and political decisions regarding flood risk management after flood hazard assessment, it is necessary to provide the establishment of flood risk management. For the flood hazard estimation to identify the use of flood risk analysis as a part of an integrated flood risk management, future research should continue to provide a methodological and applied contribution to the field of flood risk analysis, which will be used for risk comparison, risk communication, setting priorities for public funding, project optimization, the elaboration of risk sharing options, and for the cost-benefit analysis. The methods and algorithms used in this work will be applicable for other watersheds with similar geomorphological and climatic conditions. In order to continue the research, the focus will be on the tasks of a quantitative risk analysis and management as:

(a) Loss analysis.
(b) Risk quantification.
(c) Evaluation of uncertainty.

Flood "loss analysis" covers a social, economic, and environmental dimension. A specific issue in flood loss analysis is the limitation of methods and data, such as loss functions and spatial information on the elements at risk. [1, 2] and [3] named a number of flood characteristics that determine the magnitude of losses. These characteristics cover the inundation depth, the flow velocity, the inundation duration, etc.

A. Ramesh, *Response of Flood Events to Land Use and Climate Change*,
Springer Theses, DOI: 10.1007/978-94-007-5527-7_8,

"Risk quantification" provides a basis for risk assessment and risk management. Risk in the context of this work is defined as a combination of the consequences of a flood event and its probability of occurrence [4]. The consequences may arise in a social, economic, and environmental dimension and may therefore affect individuals or the society [5–7]. The risk analysis may anticipate potential future events (ex-ante) or may focus on historic events (ex-post) [8].

There are manifold "uncertainties" associated with the analysis of flood risks. [9] and [10] therefore preferred the term "risk estimation" instead of "risk calculation". [11] suggested to not consider the results of risk analysis models as facts, since much uncertainty is inherent in the modeling results. There is uncertainty in the occurrence and the magnitude of extremes, which is specified by a probability density function. Then, there is uncertainty due to the lack of knowledge and the imperfect system of description, which affects the quantification of the hazard, and the consequences and probability density function.

8.2 Web-GIS Platform

This research aims to bridge the gap between "science" and "stakeholders": From a "variety of complex and complicated equations which govern on flood hazard/ risk assessment that includes a wide range of uncertainties in both temporal and spatial scales" to "easy accessible achieved results and information for both academic staff and also non-academic staff such as (non)governmental organizations and stakeholders to do any future flood risk management to reduce the vulnerability of flood based on different scenarios or modules". To this purpose visualizing the results, Web-GIS platform is necessary. Moreover, and as a consequence of this research, developing a common user-friendly Web-GIS tool for technical staff and decision makers who normally do not have a background in Geoinformatics for flood risk analysis and management, it will be a suitable solution in any future flood risk management [12].

A "Web-GIS" is a GIS platform for sharing spatial and geographical data using the Web [12]. The traditional stand-alone GIS tools are fixed and gathered by the client from the Web, removing every need for software installation and setup [13, 14]. A Web-based system as a user-friendly visualization instrument should, however, be used to facilitate interpretation, generation of basic knowledge, and representation of results, both to increase the purpose of training and to define practical risk management plans [12, 15].

8.3 Application of Remote Sensing-Based Data

Considering the lack of snow data in the Barcelonnette area, to study the effects of the extent and variability of seasonal snow cover on the climate system, future research needs to acquire snow cover information both on a fine temporal scale and in local areas. Compared to the conventional snow measurements, satellite remote sensing data are particularly well-adapted to monitoring snow-covered surfaces over continuous space–time scales [16]. Additionally, to simulate and forecast daily stream flow, it is recommended to apply the Snowmelt-Runoff Model (SRM).

8.4 Application of GCMs and RCMs Data

In current research, historical investigation into climate parameters confirmed some sign of climate change in the area. To continue this part of the research, it is necessary to estimate the global scenarios for climate change. The Intergovernmental panel on climate change (IPCC) developed global scenarios which contain various driving forces of climate change, including population growth and socio-economic development [17]. These forces encompass various future scenarios that might influence greenhouse gas sources and sinks, such as the energy system and land use change. To apply these scenarios on a regional (France) and local scale (the Barcelonnette area), downscaling techniques, both dynamic and statistical, are necessary to derive future climatic parameters from both general circulation models (GCMs) and regional circulation models (RCMs). The downscaling of IPCC scenarios refers to the process of taking global information about climate response to changing atmospheric composition, and translating it into a finer spatial scale that is more meaningful in the context of local and regional impacts.

References

1. Apel H, Thieken A, Merz B, Blöschl G (2004) Flood risk assessment and associated uncertainty. Nat Hazards Earth Syst Sci 4:295–308
2. Apel H, Merz B, Thieken AH (2009) Influence of dyke breaches on flood frequency estimation. Comput Geosci 35(5):907–923
3. Messner F, Penning-Rowsell E, Green C, Meyer V, Tunstall S, van der Veen A (2006) Guidelines for socio-economic flood damage evaluation. Floodsite Project Report T9-06-01
4. Ale BJM (2002) Risk assessment practices in the Netherlands. Saf Sci 40:105–126
5. Borter P (1999) Risikoanalysen bei gravitativen naturgefahren–107/I–methode/bundesamt für umwelt, wald und landschaft (BUWAL). Bern, Switzerland
6. Merz B, Disse M, Günther K, Schumann A (2010) Preface outcomes of the RIMAX programme: risk management of extreme flood events. Natural Hazards and Earth System Sciences (NHESS) Special Issue, pp 1–3
7. Plate EJ (2002) Flood risk and flood management. J Hydrol 267:2–11
8. Vrouwenvelder T, Lovegrove R, Holicky M, Tanner P, Canisius G (2001) Risk assessment and risk communication in civil engineering, international conference of safety, Risk and Reliability Trends in Engineering, Malta, pp 885–890, 21–23 March 2001
9. Fritzsche AF (1986) Wie sicher leben wir? Risikobeurteilung und—bewältigung in unserer gesellschaft. Köln, TÜV Rheinland

10. Merz B (2006) Hochwasserrisiken: grenzen und möglichkeiten der risikoabschätzung. Schweizerbart, Stuttgart
11. Petak WJ, Atkisson AA (1982) Natural hazard risk assessment and public policy, anticipating the unexpected (springer series on environmental management). Springer, New York
12. Frigerio S, van Westen CJ (2010) Riskcity and webriskcity: data collection, display and dissemination in a multi—risk training package. Cartography geogr inf sci 37:119–135
13. Brabhaharan P, Fleming MJ, Lynch R (2001) Natural hazard risk management for road networks, Part I: Risk management strategies, research report 217. Transfund New Zealand, New Zealand, p 75
14. Salvati P, Calducci V, Bianchi C, Guzzetti F, Tonelli G (2009) A WebGIS for the dissemination of information on historical landslides and floods in umbria. Italy GeoInformatica 13:305–322
15. Romang H, Binder P, Spreafico M (2009) Common information platform for natural hazards in Switzerland, geophysical research abstracts. European geosciences union (EGU), Vienna
16. Pivot F, Duguay CR, Brown RD (2002). Remore sensing of snow cover for climate change monitoring in the Canadian Subarctic: A comparison between SMMR-SSM/I and NOAA-AVAHRR secsors, 59th Eastern snow conference, Stowe, Vermont, USA
17. Environmental Software & Services GmbH (ESS), Available online at: http://www.ess.co.at/METEO/downscaling.html (Verified 28 Nov 2011)

Summary

Scientific, political and public opinion is agreed that we live in a period of accelerated flood hazard. Flood events are among the most common, costly and deadly natural hazards. Throughout history they have been a major concern for people residing near rivers and coastal areas. River floods may occur whenever the capacity of the channel or manmade drainage system is unable to cope with the volume of water generated by rainfall or when flood defenses fail. Since technical flood protection systems have limited resistance, the sensitive points in protected areas are exposed to a residual flood hazard. In order to have available an effective informational tool, as well as a valuable basis for setting priorities on technical, financial and political decisions regarding flood hazard/risk management, it is necessary to provide/create flood hazard maps and analyze the triggering factors which influence floods.

This study focused on the floodplain of the Ubaye River in the Barcelonnette area, which is located in the Alpes de Haute Provence, southeast of France. The Barcelonnette area is situated in an elongated expanse, extending below mountain slopes and along a river channel. In case of overtopping or dyke failure, the whole area will be impacted by a flood. The Barcelonnette basin is affected by a Mediterranean marine influence and moderated by a mild continental mountain climate.

In this research, the system description, hazard identification, and hazard analysis were estimated by computing of a set of scenarios. The variability of flood crisis scenarios (e.g. dyke failure) was estimated for civil protection purposes in the study area, with respect to different DEM resolutions, surface elements which affect water flow and a variety of cross sections and river morphology over time. A selection of a sub-set of these scenarios with distinct return periods was applied as input information into the hydraulic (1D) & hydrologic (1D–2D) models. In principle, the analysis of flood hazard was based on two approaches: an anticipative study of potential future events and the investigation of events that

A. Ramesh, *Response of Flood Events to Land Use and Climate Change*,
Springer Theses, DOI: 10.1007/978-94-007-5527-7,

already occurred. The results of the hydrodynamic section are very important as basic information for policy making, decision support and flood hazard planning in the Barcelonnette area. Additionally, this research gave a comparative overview of the major challenges faced when dealing with flood hazards. It provided a methodology to investigate the probable climate change and land use change from the past until the present in the Barcelonnette area. The results of the time series analysis between 1961 and 2009 showed that the highest precipitation occurred in 1977. Therefore, this year was selected for comparison with the 2008 flood. Monthly data for precipitation discharge and temperature was analyzed for both years. By investigating all achieved results from the time series analysis, it could be concluded that: (a) the amount of precipitation for almost all months in 1977 was more than in 2008. In general, precipitation shows a decreasing trend at the end of the period (1961–2008); (b) the temperature for almost all months in 2008 increased from 0–4 °C; (c) maximum discharge in 2008 was more than in 1977. To confirm these results, the "Mann–Kendall" method was applied to detect possible influence of climate change in the area. From all achieved results, it could be concluded that although there is a significant downward trend in precipitation and, inversely, an upward trend in the maximum temperature, the distribution of discharge in the channel was normal and no trend was distinguished in the flow discharge at the Barcelonnette station. The analysis of average monthly precipitation between 1961 and 2009 showed that the maximum rainfall occurred in spring and autumn and the maximum peak was recorded in May, June, October and November. Based on observed data at the Barcelonnette station from 1961 to 2009, the maximum temperature occurred between May and September. During the same period, a monthly investigation into discharge showed that lower discharge happened from December to March, while the maximum amount of water occurred from April to July. These results pushed the research to find other triggering factors which affect on normal discharge distribution in the channel. In light of this, the available daily basis analysis for maximum snow depth, maximum discharge, precipitation and maximum temperature between 1995 and 2005 was also analyzed in this research. The results showed that during the summer the rate of snowmelt had a direct impact on the rivers' peak discharge. In both flood events in 1957 and 2008, it was also reported that the rate of snowmelt was higher than the average of other years. From the investigated results of land use change, it was realized that urban areas have increased about $\sim$+140 % and arable land has had a decreasing trend (~ -54 %) from the past until the present. Forests also represented an increasing trend (~ 12 %), inversely due to channelization and human activity, and the channel has also decreased (~ -37 %) over time. Based on all achieved results, it's concluded that: (a) a decreasing trend in precipitation and an increasing trend in temperature and seasonal changes seem to be a sign of climate change (especially since increasing temperatures lead to more snowmelt during spring); (b) the results of cross-correlation for water depth and discharge also showed that there is a strong backward negative correlation between maximum discharge and snow depth. This means that when snow packs start to melt, the melted snow will consequently affect the rate of runoff in the Ubaye River; (c) the quick rate of expanding urban areas and reduction of arable lands affect the amount of discharge. This leads to more rainfall that directly flows into the channel instead of infiltrating into the soil; (d) on the other hand, reconstructing and increasing the dyke height after the severe flood in 1957 was a good strategy to protect the area from severe floods and could be considered a constructive human activity, but there are still concerns about overtopping and inundation in case of a severe flood event with the current dyke height; (e) as a consequence, it could be assumed that climate change, land use changes, anthropogenic activities and snowmelt are the most important influential factors on the discharge and flow regime fluctuation in the Ubaye River as well. Based on the results of this research, some recommendations are drawn. In addition some suggestions are also put forth for future research.

Appendix (1)

Abstract in English This research aimed to develop an available effective tool for information, as well as valuable basis for priority setting and further technical, financial and political decisions regarding flood hazard assessment. This study focused on the floodplain of the Ubaye River in the Barcelonnette area, located in the Alpes de Haute Provence, southeast of France. Based on the objectives of this research, at the first stage, the results of interpolation methods were discussed. The main aim was to find the interpolation methods, which give the best results based on statistical, spatial and hydrological analysis. Then, the system description and hazard identification have been analyzed by computation of a set of the variability of flood crisis scenarios for any civil protection purposes (e.g. dyke failure). Selection of a sub-set of these scenarios with distinct return periods as inputs into the hydraulic (1D) and hydrologic (1D–2D) models were also estimated. Additionally, this research gave a comparative overview of the major challenges faced when dealing with flood hazard. Therefore, the next stage of this research aimed to provide a methodology to investigate the probable climate change. Therefore, the non parametric Mann–Kendall method was applied in order to detect trend and to find probable abrupt points in spatial distribution of the observed time series (climatological and hydrological parameters). Based on achieved results, a decreasing trend in precipitation and an increasing trend in temperature and seasonal changes seemed to be a sign of climate change, especially since increasing temperature leads to more snowmelt during spring. In sequence, the results of cross-correlation for water depth and discharge also showed that there was a strong backward negative correlation between maximum discharge and snow depth. The obtained results of land use change detection analysis represented that the quick rate of expanding urban areas and reduction of arable lands influenced on the amount of discharge in Ubaye River. As overall consequence of this research, it could be concluded that the melting of snow, land use changes, anthropogenic activities and climate change are the most important forcing and triggering factors in the fluctuation and changes in the discharge and

A. Ramesh, *Response of Flood Events to Land Use and Climate Change*,
Springer Theses, DOI: 10.1007/978-94-007-5527-7,

flow regime of Ubaye River in Barcelonnette area. The results of this research are very important as basic information for policymaking, decision support and flood hazard planning in Barcelonnette area.

Keywords Flood hazard assessment, Trend analysis, Climate change, Land use change.

Appendix (2)

Kurzfassung Diese Forschung soll sowohl zur Entwicklung eines effektiven Informationssystems betragen, als auch eine wertvolle Grundlage für die Festlegung von Prioritäten und weiteren technischen, finanziellen und politischen Entscheidungen in Bezug auf Hochwassergefährdungsbeurteilung bilden. Der Fokus dieser Studie liegt auf dem Überschwemmungsgebiet des Ubaye Flusses im Arrondissement Barcelonnette, im Department Alpes-de-Haute-Provence, im Südosten von Frankreich. Zu Beginn der Arbeit werden Ergebnisse unterschiedlicher Interpolationsmethoden diskutiert. Das Hauptziel war es, die Interpolationsmethoden zu finden, die statistische, räumliche und hydrologische Aspekte am besten wiedergeben. Anschließend wurde das System und die Gefahrenerkennung durch Berechnung von mehreren Hochwasserszenarien für Katastrophenschutzzwecke (z.B. Deichbrüche) charakterisiert. Es wurde eine Untermenge von Szenarien mit unterschiedlichen Wiederkehrperioden als Eingang für hydraulische (1D) und hydrologische (1D–2D) Modelle ausgewählt. Darüber hinaus gab diese Forschung einen Überblick über die wichtigsten Herausforderungen im Umgang mit Hochwassergefahren. Mithilfe dieser Grundlagen soll eine Methode entwickelt werden, die die hydrologischen Auswirkungen des Klimawandels beschreibt. Es wurde die nicht parametrische Mann–Kendall-Methode angewandt, um Trends zu erkennen und Brüche in der räumlichen Verteilung der beobachteten Zeitreihen (klimatologischer und hydrologischer Parameter) zu finden. Es wurde im Untersuchungsgebiet ein abnehmender Trend der Niederschläge und ein zunehmender Trend in der Temperatur und den jahreszeitlichen Schwankungen vorhergesagt, besonders zu beachten ist die steigende Temperatur, die zu mehr Schneeschmelze im Frühjahr führt. Aufgrund der Ergebnisse der Kreuzkorrelation zwischen Wassertiefe und Abfluss zeigte sich auch eine starke negative Korrelation zwischen maximalem Abfluss und Schneehöhe. Die Analyse der Landnutzungsänderungen ergab, dass die schnell wachsenden urbanen Gebieten und die Verringerung der Anbauflächen den Abluss des Flusses Ubaye beeinflussen. Allgemein konnte gezeigt werden, dass das

A. Ramesh, *Response of Flood Events to Land Use and Climate Change*, Springer Theses, DOI: 10.1007/978-94-007-5527-7,

Schmelzen von Schnee, Landnutzungsänderungen, anthropogenen Aktivitäten und der Klimawandel die wichtigsten Einflussfaktoren auf die Fluktuation und Änderungen des Abflussregimes des Ubaye Flusses im Arrondissement Barcelonnette sind. Die Ergebnisse dieser Forschungen sind sehr wichtig, da grundlegende Informationen für die Politikgestaltung, Entscheidungshilfen und Hochwassergefahrenkarten zur Planung im Arrondissement Barcelonnette zur Verfügung gestellt werden.

Schlüsselwörtern Hochwasser, Gefährdungsbeurteilung, Trendanalysen, Klimawandel, Landnutzungsänderungen.

Appendix (3)

Acronyms and Abbreviations

1D	One dimensional
1D–2D	One–two dimensional
ANUDEM	Australian National University Digital Elevation Model
CRED	Centre for Research on the Epidemiology of Disasters
DEM	Digital elevation model
EEA	European Environment Agency
EM-DAT	International disaster database
GIS	Geographic information system
HEC-RAS	Hydrologic engineering centers river analysis system
IDW	Inverse distance weighted
ILWIS	Integrated land and water information system
IPCC	Intergovernmental panel on climate change
LiDAR	Light detection and ranging
OMIV	Observatoire Multidisciplinaire des Instabilités de Versants
PER	Plan d'Exposition aux Risques
PPR	Prévention des Risques Naturels—Plan for Prevention of Predictable Natural Risks
RADAR	Radio detection and ranging
RMSE	Root mean square error
RTM	Service de Restauration des Terrains de Montage—Mountain and Restoration Service
SOBEK	Hydrological model, model developed by delft hydraulics in The Netherlands
TIN	Triangular irregular network
WMO	World Meteorlogical Organizaion

A. Ramesh, *Response of Flood Events to Land Use and Climate Change*,
Springer Theses, DOI: 10.1007/978-94-007-5527-7,

Appendix (4)

Additional Information

Figs. A.1, A.2, A.3, A.4, A.5, A.6, A.7

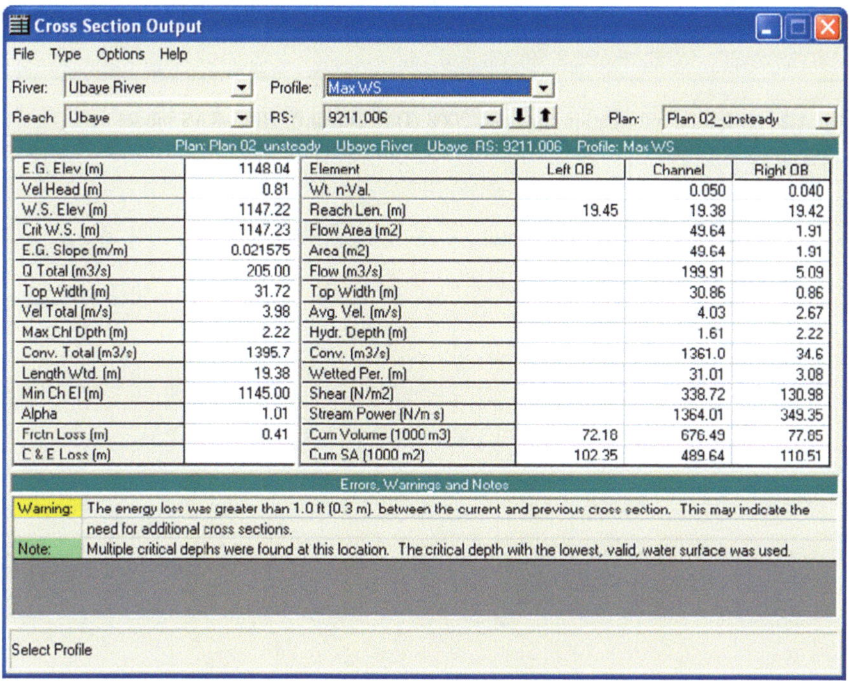

Fig. A.1 Hydraulic information for flood 2008 (Upstream), HEC-RAS model

A. Ramesh, *Response of Flood Events to Land Use and Climate Change*,
Springer Theses, DOI: 10.1007/978-94-007-5527-7,

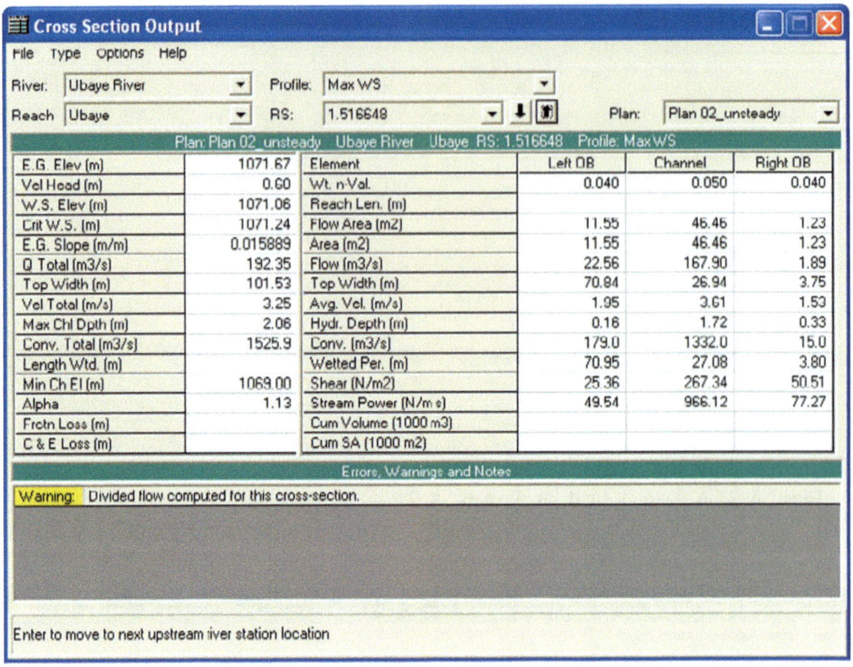

E.G. Elev (m)	1071.67
Vel Head (m)	0.60
W.S. Elev (m)	1071.06
Crit W.S. (m)	1071.24
E.G. Slope (m/m)	0.015889
Q Total (m3/s)	192.35
Top Width (m)	101.53
Vel Total (m/s)	3.25
Max Chl Dpth (m)	2.06
Conv. Total (m3/s)	1525.9
Length Wtd. (m)	
Min Ch El (m)	1069.00
Alpha	1.13
Frctn Loss (m)	
C & E Loss (m)	

Element	Left OB	Channel	Right OB
Wt. n-Val.	0.040	0.050	0.040
Reach Len. (m)			
Flow Area (m2)	11.55	46.46	1.23
Area (m2)	11.55	46.46	1.23
Flow (m3/s)	22.56	167.90	1.89
Top Width (m)	70.84	26.94	3.75
Avg. Vel. (m/s)	1.95	3.61	1.53
Hydr. Depth (m)	0.16	1.72	0.33
Conv. (m3/s)	179.0	1332.0	15.0
Wetted Per. (m)	70.95	27.08	3.80
Shear (N/m2)	25.36	267.34	50.51
Stream Power (N/m s)	49.54	966.12	77.27
Cum Volume (1000 m3)			
Cum SA (1000 m2)			

Fig. A.2 Hydraulic information for flood 2008 (Downstream), HEC–RAS model

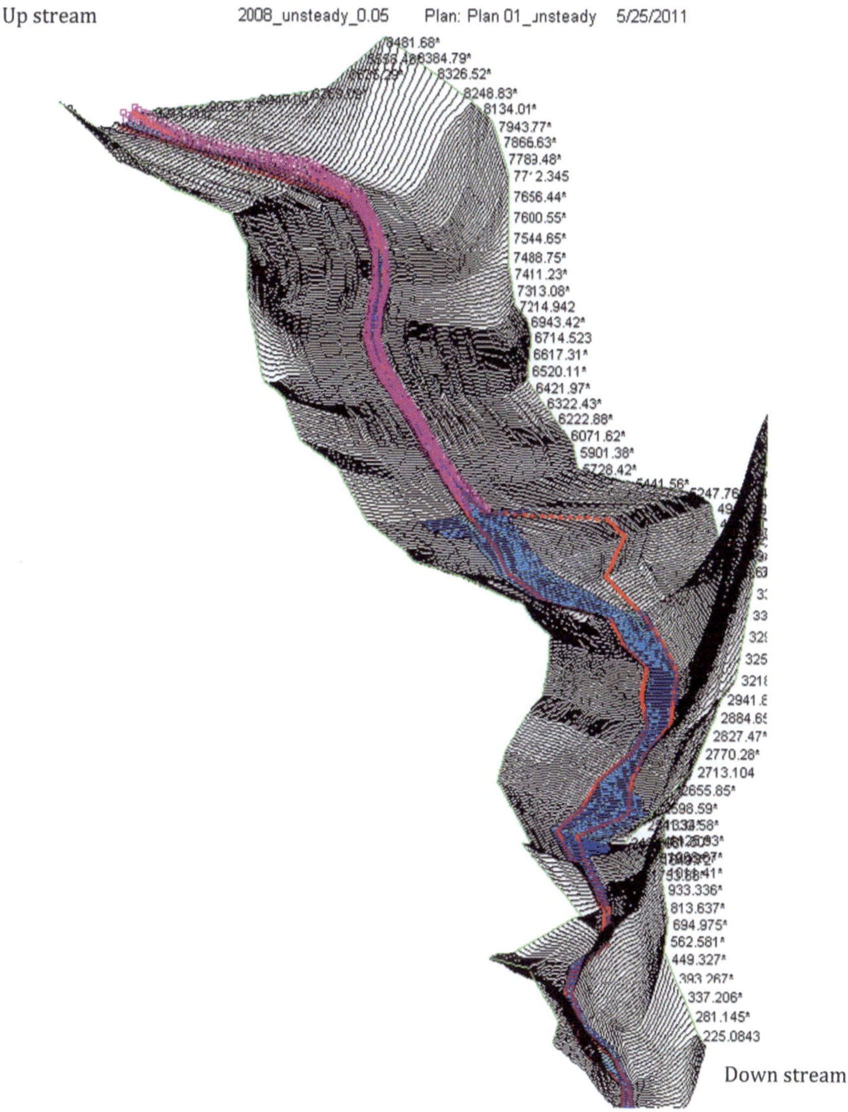

Fig. A.3 Ubaye River profile, HEC–RAS model

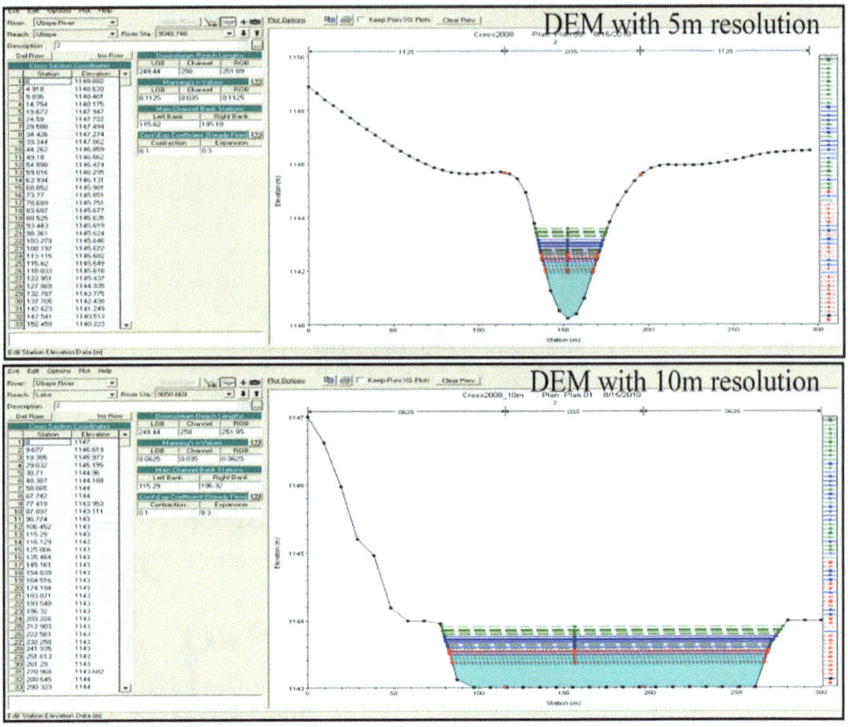

Fig. A.4 Setting up cross section, roughness and height to the dyke in HEC–RAS model

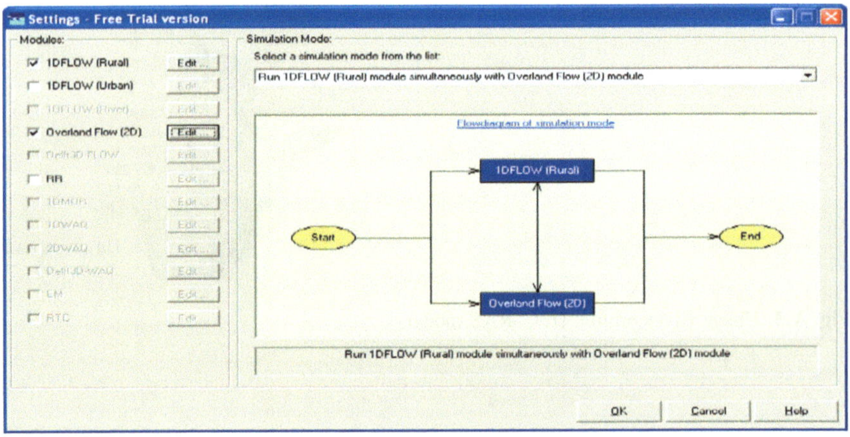

Fig. A.5 1D–2D setting in SOBEK

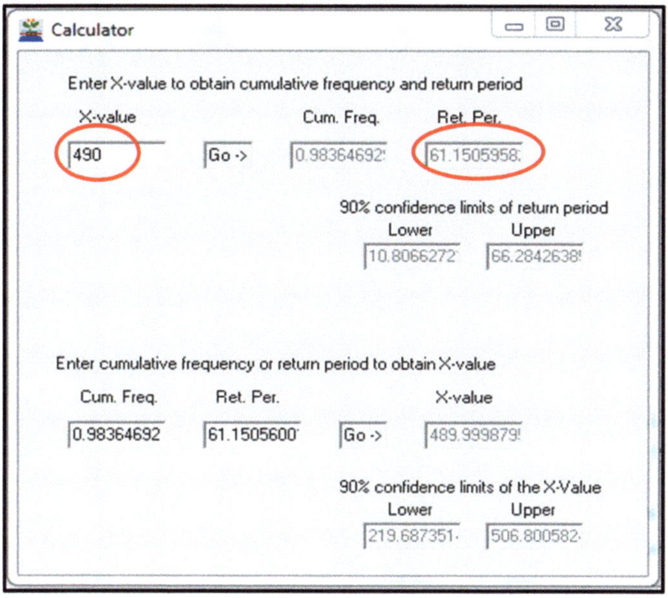

Fig. A.6 The applied model to calculate the return period frequency

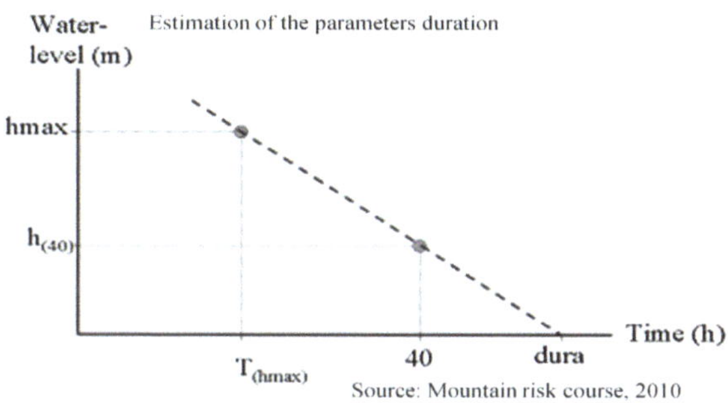

Fig. A.7 Estimation of the parameters duration

MIX
Papier aus verantwortungsvollen Quellen
Paper from responsible sources
FSC® C105338

If you have any concerns about our products,
you can contact us on
ProductSafety@springernature.com

In case Publisher is established outside the EU,
the EU authorized representative is:
Springer Nature Customer Service Center GmbH
Europaplatz 3, 69115 Heidelberg, Germany

Printed by Libri Plureos GmbH
in Hamburg, Germany